T0281571

Advanced Topics in Applied Mathematics

This book is ideal for engineering, physical science, and applied mathematics students and professionals who want to enhance their mathematical knowledge. *Advanced Topics in Applied Mathematics* covers four essential applied mathematics topics: Green's functions, integral equations, Fourier transforms, and Laplace transforms. Also included is a useful discussion of topics such as the Wiener-Hopf method, finite Hilbert transforms, Cagniard–De Hoop method, and the proper orthogonal decomposition. This book reflects Sudhakar Nair's long classroom experience and includes numerous examples of differential and integral equations from engineering and physics to illustrate the solution procedures. The text includes exercise sets at the end of each chapter and a solutions manual, which is available for instructors.

Sudhakar Nair is the Associate Dean for Academic Affairs of the Graduate College, Professor of Mechanical Engineering and Aerospace Engineering, and Professor of Applied Mathematics at the Illinois Institute of Technology in Chicago. He is a Fellow of the ASME, an Associate Fellow of the AIAA, and a member of the American Academy of Mechanics as well as Tau Beta Pi and Sigma Xi. Professor Nair is the author of numerous research articles and *Introduction to Continuum Mechanics* (2009).

ADVANCED TOPICS IN APPLIED MATHEMATICS

For Engineering and the Physical Sciences

Sudhakar Nair

Illinois Institute of Technology

CAMBRIDGE
UNIVERSITY PRESS

CAMBRIDGE
UNIVERSITY PRESS

32 Avenue of the Americas, New York NY 10013-2473, USA

Cambridge University Press is part of the University of Cambridge.

It furthers the University's mission by disseminating knowledge in the pursuit of
education, learning and research at the highest international levels of excellence.

www.cambridge.org
Information on this title: www.cambridge.org/9781107448759

© Sudhakar Nair 2011

First published 2011
First paperback edition 2014

A catalogue record for this publication is available from the British Library

Library of Congress Cataloguing in Publication data
Nair, Sudhakar, 1944– author.
Advanced Topics in Applied Mathematics: for Engineering and the Physical
Sciences/Sudhakar Nair.
p.   cm
Includes index.
ISBN 978-1-107-00620-1 (hardback)
1. Differential equations.   2. Engineering mathematics.   3. Mathematical physics.
I. Title.
TA347.D45N35 2011
620.001'51–dc22   2010052380

ISBN 978-1-107-00620-1 Hardback
ISBN 978-1-107-44875-9 Paperback

Contents

v

Preface

This text is aimed at graduate students in engineering, physics, and applied mathematics. I have included four essential topics: Green's functions, integral equations, Fourier transforms, and Laplace transforms. As background material for understanding these topics, a course in complex variables with contour integration and analytic continuation and a second course in differential equations are assumed. One may point out that these topics are not all that advanced – the expected advanced-level knowledge of complex variables and a familiarity with the classical partial differential equations of physics may be used as a justification for the term "advanced." Most graduate students in engineering satisfy these prerequisites. Another aspect of this book that makes it "advanced" is the expected maturity of the students to handle the fast pace of the course. The fours topics covered in this book can be used for a one-semester course, as is done at the Illinois Institute of Technology (IIT). As an application-oriented course, I have included techniques with a number of examples at the expense of rigor. Materials for further reading are included to help students further their understanding in special areas of individual interest. With the advent of multiphysics computational software, the study of classical methods is in general on a decline, and this book is an attempt to optimize the time allotted in the curricula for applied mathematics.

I have included a selection of exercises at the end of each chapter for instructors to choose as weekly assignments. A solutions manual for these exercises is available on request. The problems are numbered in such a way as to simplify the assignment process, instead of clustering a number of similar problems under one number.

Classical books on integral transforms by Sneddon and on mathematical methods by Morse and Feshbach and by Courant and Hilbert form the foundation for this book. I have included sections on the Boundary Element Method and Proper Orthogonal Decomposition under integral equations – topics of interest to the current research community. The Cagniard–De Hoop method for inverting combined Fourier-Laplace transforms is well known to researchers in the area of elastic waves, and I feel it deserves exposure to applied mathematicians in general. Discrete Fourier transform leading to the fast Fourier algorithm and the Z-transform are included.

I am grateful to my numerous students who have read my notes and corrected me over the years. My thanks also go to my colleagues, who helped to proofread the manuscript, Kevin Cassel, Dietmar Rempfer, Warren Edelstein, Fred Hickernell, Jeff Duan, and Greg Fasshauer, who have been persistent in instilling applied mathematics to believers and nonbelievers at IIT, and, especially, for training the students who take my course. I am also indebted to my late colleague, Professor L. N. Tao, who shared the applied mathematics teaching with me for more than twenty-five years.

The editorial assistance provided by Peter Gordon and Sara Black is appreciated.

The Mathematica™ package from Wolfram Research was used to generate the number function plots.

My wife, Celeste, has provided constant encouragement throughout the preparation of the manuscript, and I am always thankful to her.

GREEN'S FUNCTIONS

Before we introduce the **Green's functions**, it is necessary to familiarize ourselves with the idea of **generalized functions** or **distributions**. These are called generalized functions as they do not conform to the definition of functions. They are often unbounded and discontinuous. They are characterized by their integral properties as linear functionals.

1.1 HEAVISIDE STEP FUNCTION

Although this is a simple discontinuous function (not a generalized function), the **Heaviside step function** is a good starting point to introduce generalized functions. It is defined as

$$h(x) = \begin{cases} 0, & x < 0, \\ 1/2, & x = 0, \\ 1, & x > 0. \end{cases} \tag{1.1}$$

The value of the function at $x = 0$ is seldom needed as we always approach the point $x = 0$ either from the right or from the left (see Fig. 1.1). When we consider representation of this function using, say, Fourier series, the series converges to the mean of the right and left limits if there is a discontinuity. Thus, $h(0) = 1/2$ will be the converged result for such a series.

Using the Heaviside function, we can express the **signum** function, which has a value of 1 when the argument is positive, and a value of -1

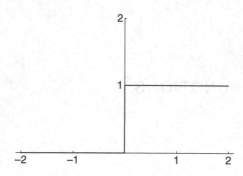

Figure 1.1. Heaviside step function.

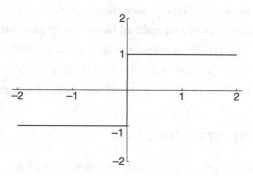

Figure 1.2. Signum function sgn(x).

when the argument is negative (see Fig. 1.2), as

$$\text{sgn}(x) = 2h(x) - 1. \tag{1.2}$$

We may convert an even function of x to an odd function simply by multiplying by sgn(x).

The function shown in Fig. 1.3 can be written as

$$f(x) = h(a - |x|). \tag{1.3}$$

This is known as the **Haar function**, which plays an important role in image processing as a basis for **wavelet** expansions. In wavelet analysis,

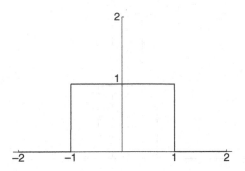

Figure 1.3. Haar function ($a = 1$).

families of Haar functions with support $a, a/2, a/4, \ldots, a/2^n$ are used as a basis to represent functions.

1.2 DIRAC DELTA FUNCTION

The Dirac delta function has its origin in the idea of *concentrated* charges in electromagnetics and quantum mechanics. In mechanics, the Dirac delta $\delta(x)$ is useful in representing concentrated forces. We can view this generalized function as the derivative of the Heaviside function, which is zero everywhere except at the origin. At the origin it is infinity. As a consequence, its integral from $-\epsilon$ to $+\epsilon$ is unity. As is the case for all generalized functions, we consider the delta function as the limit of various sequences of functions. For example, consider the sequence of functions shown in Fig. 1.4, which depends on the parameter ϵ,

$$f(x;\epsilon) = \begin{cases} 0, & |x| > \epsilon, \\ \frac{1}{2\epsilon}, & |x| < \epsilon. \end{cases} \qquad (1.4)$$

In the limit $\epsilon \to 0, f(x;\epsilon) \to \delta(x)$.

Note that

$$\int_{-\infty}^{\infty} f(x;\epsilon)\,dx = \int_{-\epsilon}^{\epsilon} f(x;\epsilon)\,dx = 1 = \int_{-\infty}^{\infty} \delta(x)\,dx = \int_{-\epsilon}^{\epsilon} \delta(x)\,dx. \quad (1.5)$$

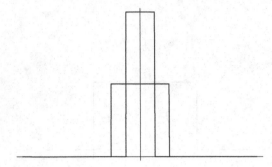

Figure 1.4. A delta sequence using Haar functions.

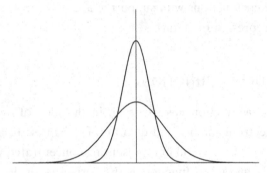

Figure 1.5. Another delta sequence using probability functions.

Another sequence of continuous functions which forms a delta sequence is given (see Fig. 1.5) by the Gauss functions or probability functions:

$$f(x;n) = \frac{n}{\sqrt{\pi}} e^{-n^2 x^2}. \tag{1.6}$$

We can see that the area under this curve remains unity for all values of n. Let

$$I = \int_{-\infty}^{\infty} n e^{-n^2 x^2}\, dx = \int_{-\infty}^{\infty} e^{-x^2}\, dx, \tag{1.7}$$

where we substituted $nx \to x$. Using polar coordinates,

$$I^2 = \int_{-\infty}^{\infty} e^{-x^2}\, dx \int_{-\infty}^{\infty} e^{-y^2}\, dy$$

$$= \int_{-\infty}^{\infty} \int_{-\infty}^{\infty} e^{-(x^2+y^2)}\, dxdy$$

$$= \int_{0}^{\infty} \int_{0}^{2\pi} e^{-r^2} r\, drd\theta$$

$$= 2\pi \int_{0}^{\infty} e^{-r^2} r\, dr = -\pi e^{-r^2} \Big|_{0}^{\infty} = \pi. \qquad (1.8)$$

We frequently encounter the integral I, which has the value

$$I = \int_{-\infty}^{\infty} e^{-x^2}\, dx = \sqrt{\pi}. \qquad (1.9)$$

As $n \to \infty$, $f(x;n) \to \delta(x)$.

By shifting the origin from $x = 0$ to $x = \xi$, we can move the spike of the delta function to the point ξ. This new function has the properties,

$$\delta(x - \xi) = 0, \quad x \neq \xi, \qquad (1.10)$$

$$\int_{\xi-\epsilon}^{\xi+\epsilon} \delta(x - \xi)\, dx = 1. \qquad (1.11)$$

An important property of the delta function is localization under integration. As usual, properties of the generalized functions are proved using the corresponding sequences. For any smooth function $\phi(x)$, which is nonzero only in a finite interval (a, b), using the sequence (1.4), we have

$$\int_{-\infty}^{\infty} \phi(x)\delta(x - \xi)\, dx$$

$$= \lim_{\epsilon \to 0} \int_{\xi-\epsilon}^{\xi+\epsilon} \phi(x) \frac{dx}{2\epsilon}$$

$$= \lim_{\epsilon \to 0} \int_{\xi-\epsilon}^{\xi+\epsilon} \left[\phi(\xi) + \phi'(\xi)(x - \xi) + \frac{1}{2}\phi''(\xi)(x - \xi)^2 + \cdots \right] \frac{dx}{2\epsilon}$$

$$= \lim_{\epsilon \to 0}\left[\phi(\xi) + \frac{1}{2}\phi'(\xi)\epsilon + \frac{1}{6}\phi''(\xi)\epsilon^2 + \cdots\right]$$

$$= \phi(\xi). \tag{1.12}$$

Integrals involving a *scaled* delta function can be evaluated as shown:

$$\int_{-\infty}^{\infty} \phi(x)\delta(\frac{x-\xi}{a})\,dx = \int_{-\infty}^{\infty} \phi(ax')\delta(x'-\xi')\,adx'$$

$$= a\phi(\xi), \tag{1.13}$$

where we used $x' = x/a$, $\xi' = \xi/a$, $a > 0$.

1.2.1 Macaulay Brackets

A simplified notation to represent integrals of the δ function was introduced in the context of structural mechanics by Macaulay. In this notation

$$\delta(x-\xi) = \langle x-\xi\rangle^{-1}, \tag{1.14}$$

$$h(x-\xi) = \langle x-\xi\rangle^{0}, \tag{1.15}$$

$$\int \langle x-\xi\rangle^n dx = \frac{1}{n+1}\langle x-\xi\rangle^{n+1}, \quad n \neq -1, \tag{1.16}$$

$$\int \langle x-\xi\rangle^{-1} dx = \langle x-\xi\rangle^{0}. \tag{1.17}$$

All of these functions are zero when the quantity inside the brackets is negative. For $n < 0$, some books omit the factor $1/(n+1)$ in the integral. We may include higher derivatives of the delta function in this group. In one-dimensional problems, such as the deflection of beams under concentrated loads, this notation is useful.

1.2.2 Higher Dimensions

In an n-dimensional Euclidian space $\mathbf{R}^n$ with coordinates $(x_1, x_2, \ldots, x_n)$, we use the simplified notation for the infinitesimal volume,

$$dx_1 dx_2 \ldots dx_n = dx, \qquad (1.18)$$

and the same for functions

$$\phi(x_1, x_2, \ldots, x_n) = \phi(x), \quad \delta(x_1, x_2, \ldots, x_n) = \delta(x). \qquad (1.19)$$

Then the n-dimensional integral,

$$\int_{\mathbf{R}^n} \phi(x)\delta(x)\, dx = \phi(0). \qquad (1.20)$$

More often we encounter situations involving two and three-dimensional spaces and cartesian coordinates (x, y) or (x, y, z), and the above result directly applies. When we use polar coordinates (or spherical coordinates) the appropriate area element (or volume element)

$$dA = r\, dr\, d\theta \quad (\text{or} \quad dV = r^2 \sin^2\phi\, dr\, d\phi\, d\theta) \qquad (1.21)$$

is used. For example,

$$\int_0^\infty \int_0^{2\pi} f(r, \theta)\delta(r - r_0, \theta - \theta_0) r\, dr\, d\theta = f(r_0, \theta_0) \qquad (1.22)$$

1.2.3 Test Functions, Linear Functionals, and Distributions

We conclude this section by introducing the idea of generalized functions or distributions as linear functionals over test functions.

A function, $\phi(x)$, is called a test function if (a) $\phi \in C^\infty$, (b) it has a closed bounded (compact) support, and (c) ϕ and all of its derivatives decrease to zero faster than any power of $|x|^{-1}$.

A linear functional $\mathcal{T}$ of ϕ maps it into a scalar. This is done using an integral over $-\infty$ to ∞ as an inner product with some other sequence or distribution, f. If we denote this mapping as

$$T_f[\phi] = \int_{-\infty}^{\infty} f(x)\phi(x)\,dx, \tag{1.23}$$

then the δ-distribution is defined by the relation

$$T_\delta[\phi] = \int_{-\infty}^{\infty} \delta(x)\phi(x)\,dx = \phi(0). \tag{1.24}$$

A sequence $\delta_n\,(n = 0, 1, \ldots, \infty)$ converges to the δ-function if

$$\lim_{n \to \infty} T_{\delta_n}[\phi] \to \phi(0). \tag{1.25}$$

A distribution $\mu(x)$ is the derivative of the δ-distribution if

$$T_\mu[\phi] = -\phi'(0), \tag{1.26}$$

as

$$\int_{-\infty}^{\infty} \delta'(x)\phi(x)\,dx = -\int_{-\infty}^{\infty} \delta(x)\phi'(x)\,dx = -\phi'(0). \tag{1.27}$$

This way, we can define higher-order derivatives of the delta function. In engineering, concentrated forces, charges, fluid flow sources, vortex lines, and the like are represented using delta functions. The delta function is also called a unit impulse function in control theory.

1.2.4 Examples: Delta Function

Using the property, for any test function ϕ,

$$\int_{-\infty}^{\infty} \phi(x)\psi(x)\,dx = \phi(\xi) \tag{1.28}$$

implies

$$\psi(x) = \delta(x - \xi), \tag{1.29}$$

prove that, for $\alpha, \beta \neq 0$,

(a)

$$\frac{\partial}{\partial \alpha}\delta(\alpha x) = -\frac{1}{\alpha^2}\delta(x), \tag{1.30}$$

(b)

$$\delta(e^{\alpha x} - \beta) = \frac{1}{\alpha\beta}\delta\left(x - \frac{\ln\beta}{\alpha}\right). \tag{1.31}$$

We may solve these examples using the basic properties of the delta function as follows:

(a)

$$\int_{-\infty}^{\infty} \phi(x)\frac{\partial}{\partial\alpha}\delta(\alpha x)\,dx = \frac{\partial}{\partial\alpha}\int_{-\infty}^{\infty}\phi(x)\delta(\alpha x)\,dx$$

$$= \frac{\partial}{\partial\alpha}\int_{-\infty}^{\infty}\phi(x/\alpha)\delta(x)\,dx/\alpha$$

$$= \frac{\partial}{\partial\alpha}\frac{1}{\alpha}\phi(0)$$

$$= -\frac{1}{\alpha^2}\phi(0). \tag{1.32}$$

Comparing

$$\frac{\partial}{\partial\alpha}\delta(\alpha x) = -\frac{1}{\alpha^2}\delta(x). \tag{1.33}$$

(b)

$$\int_{-\infty}^{\infty} \phi(x)\delta(e^{\alpha x} - \beta)\,dx = \int_{-\infty}^{\infty}\phi\left(\frac{\ln y}{\alpha}\right)\delta(y - \beta)\frac{dy}{\alpha y}$$

$$= \int_{-\infty}^{\infty}\frac{1}{\alpha y}\phi\left(\frac{\ln y}{\alpha}\right)\delta(y - \beta)\,dy$$

$$= \frac{1}{\alpha\beta}\phi\left(\frac{\ln\beta}{\alpha}\right). \tag{1.34}$$

Thus,

$$\delta(e^{\alpha x} - \beta) = \frac{1}{\alpha\beta}\delta\left(x - \frac{\ln\beta}{\alpha}\right). \tag{1.35}$$

Other well-known examples of delta sequences are

$$\delta_a = \frac{1}{\pi}\frac{a}{a^2 + x^2}, \quad \text{limit} \quad a \to 0, \tag{1.36}$$

$$\delta_\lambda = \frac{1}{\pi}\frac{\sin\lambda x}{x}, \quad \text{limit} \quad \lambda \to \infty. \tag{1.37}$$

1.3 LINEAR DIFFERENTIAL OPERATORS

Consider the differential equation

$$Lu(x) = f(x); \quad a < x < b. \tag{1.38}$$

Here, $u(x)$ is the unknown, $f(x)$ is a given forcing function, and L is a differential operator. For a differential equation of order n, we need n boundary conditions. For the time being, let us assume all the needed boundary conditions are homogeneous. The differential operator L has the form

$$L = a_n(x)\frac{d^n}{dx^n} + a_{n-1}(x)\frac{d^{n-1}}{dx^{n-1}} + \cdots + a_0(x). \tag{1.39}$$

A linear operator satisfies the properties

$$L(u_1 + u_2) = Lu_1 + Lu_2, \tag{1.40}$$

$$L(cu) = cLu, \tag{1.41}$$

where u_1, u_2, and u are functions in $\mathcal{C}^n$, and c is a constant. Recall $\mathcal{C}^n$ indicates the set of differentiable functions with all derivatives up to and including the nth continuous.

1.3.1 Example: Boundary Conditions

For the system

$$\frac{d^2u}{dx^2} + u = \sin x; \quad u(1) = 1, \quad u(2) = 3, \tag{1.42}$$

with nonhomogeneous boundary conditions, we introduce a new dependent variable v, as

$$u = v + Ax + B. \tag{1.43}$$

The boundary conditions become

$$u(1) = 1 = v(1) + A + B, \tag{1.44}$$

$$u(2) = 3 = v(2) + 2A + B. \tag{1.45}$$

We get homogeneous boundary conditions for v, if we choose

$$A + B = 1, \tag{1.46}$$

$$2A + B = 3, \tag{1.47}$$

with the solutions, $A = 2$ and $B = -1$. The original differential equation becomes

$$\frac{d^2 v}{dx^2} + v = \sin x - 2x + 1; \quad v(1) = 0, \quad v(2) = 0. \tag{1.48}$$

1.4 INNER PRODUCT AND NORM

Given two real functions $u(x)$ and $v(x)$ on $x \in (a,b)$, we define their inner product as

$$\langle u, v \rangle = \int_a^b u(x)v(x)\,dx. \tag{1.49}$$

The two functions u and v are orthogonal if

$$\langle u, v \rangle = 0. \tag{1.50}$$

The Euclidian norm of a function u associated with the above inner product is defined as

$$\|u\| = \sqrt{\langle u, u \rangle} = \left\{ \int_a^b u^2\,dx \right\}^{1/2}. \tag{1.51}$$

If the norm of a function is unity, we call it a normalized function. For any function, by dividing it by its norm, we obtain a new normalized version of the function. Generalized inner product and norms are defined by inserting a weight function w, which is positive, in the definitions, as

$$\langle u, v \rangle = \int_a^b uvw\,dx; \quad w(x) > 0. \tag{1.52}$$

A sequence of functions, $u_i, i = 1, 2, \ldots, n$, is called an ortho-normal sequence if

$$\langle u_i, u_j \rangle = \delta_{ij}, \tag{1.53}$$

where δ_{ij} is the **Kronecker delta**, defined by

$$\delta_{ij} = \begin{cases} 1, & i = j, \\ 0, & i \neq j. \end{cases} \tag{1.54}$$

1.5 GREEN'S OPERATOR AND GREEN'S FUNCTION

For linear systems of equations in the matrix form,

$$Ax = y, \tag{1.55}$$

if A is non-singular, we can write the solution x as

$$x = By; \quad B = A^{-1}. \tag{1.56}$$

Here B is the inverse of the operator A. For the differential equation

$$Lu(x) = f(x), \tag{1.57}$$

formally, we can write the solution u as

$$u = L^{-1}f. \tag{1.58}$$

Since L is a differential operator, we expect its inverse to be an integral operator. Thus,

$$u = Gf, \quad G = L^{-1}, \tag{1.59}$$

where G is called the Green's operator. Explicitly, we have

$$u(x) = \int_a^b g(x,\xi)f(\xi)\,d\xi. \tag{1.60}$$

The kernel inside the integral, $g(x,\xi)$, is called the Green's function for the operator L.

The Green's function depends on the differential operator and the boundary conditions. Once $g(x,\xi)$ is obtained, solutions can be generated by entering the function $f(x)$ inside the integral. Thus, the task of obtaining complementary and particular solutions of differential equations for specific forcing functions becomes much simpler. Even discontinuous forcing functions can be accommodated inside the integral.

1.5.1 Examples: Direct Integrations

For the first-order differential equation

$$\frac{du}{dx} = f(x); \quad u(a) = 0, \tag{1.61}$$

by integrating we find

$$u(x) = \int_a^x f(\xi)\, d\xi. \tag{1.62}$$

For this case,

$$g(x,\xi) = h(x - \xi) = \begin{cases} 1, & \xi < x, \\ 0, & \xi > x. \end{cases} \tag{1.63}$$

The second-order differential equation describing the deflection of a taut string,

$$\frac{d^2 u}{dx^2} = f(x); \quad u(a) = 0; \quad u(b) = 0, \tag{1.64}$$

after one integration gives

$$\frac{du}{dx} = C + \int_a^x f(\xi)\, d\xi, \tag{1.65}$$

where C has to be found. Integrating again,

$$u(x) = Cx + D + \int_{x'=a}^{x'=x} \int_{\xi=a}^{\xi=x'} f(\xi)\, d\xi\, dx'. \tag{1.66}$$

Using $u(a) = 0$ we get $D = -Ca$. As shown in Fig. 1.6, the double integration has to be performed on the upper triangular area. By interchanging x'- and ξ-integrations, we have

$$u(x) = C(x - a) + \int_{\xi=a}^{\xi=x} \int_{x'=\xi}^{x'=x} f(\xi)\, dx'\, d\xi. \tag{1.67}$$

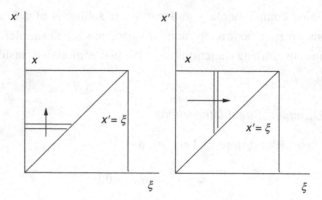

Figure 1.6. Area of integration in the x', ξ–plane.

This interchange of integrations with appropriate changes in the limits is called Fubini's theorem. After completing the x'-integration, we get

$$u(x) = C(x-a) + \int_a^x (x-\xi) f(\xi) \, d\xi. \qquad (1.68)$$

Using $u(b) = 0$, we solve for C,

$$C = -\frac{1}{b-a} \int_a^b (b-\xi) f(\xi) \, d\xi. \qquad (1.69)$$

Now u has the form

$$u = \frac{1}{b-a} \left\{ \int_a^x (b-a)(x-\xi) f(\xi) \, d\xi - \int_a^b (x-a)(b-\xi) f(\xi) \, d\xi \right\}.$$

In the second integral the range of integration, a to b, can be split into a to x and x to b.

$$u = -\frac{1}{b-a} \left\{ \int_a^x (b-x)(\xi-a) f(\xi) \, d\xi + \int_x^b (x-a)(b-\xi) f(\xi) \, d\xi \right\}.$$

We can extract the Green's function in the form

$$g(x,\xi) = \frac{1}{b-a} \begin{cases} (x-b)(\xi-a), & \xi < x, \\ (\xi-b)(x-a), & \xi > x. \end{cases} \qquad (1.70)$$

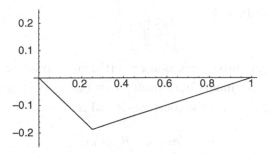

Figure 1.7. Green's function for the point-loaded string when $\xi = 0.25$.

Observe the symmetry, $g(x,\xi) = g(\xi,x)$, for this case. The differential equation represents the deflection of a string under tension subjected to a distributed vertical load, $f(x)$. With $a = 0$ and $b = 1$, we have plotted the Green's function in Fig. 1.7 for $\xi = 0.25$.

Figure 1.7 shows the deflection of the string under a concentrated unit load at $x = \xi$. The Green's function $g(x,\xi)$ satisfies

$$\frac{d^2 g(x,\xi)}{dx^2} = \delta(x - \xi), \quad g(0,\xi) = g(1,\xi) = 0. \tag{1.71}$$

With ξ as a parameter, and the *delta*-function zero for $x < \xi$ and for $x > \xi$, we can solve Eq. (1.71) in two parts. If u_1 and u_2 represent the left and right solutions which satisfy homogeneous equations, we can satisfy the left boundary condition by choosing the constants of integration to have $u_1(0) = 0$. Similarly, we can choose u_2 to have $u_2(1) = 0$.

$$g(x,\xi) = A u_1(x) = Ax, \quad x < \xi, \tag{1.72}$$

$$g(x,\xi) = B u_2(x) = B(1 - x), \quad x > \xi. \tag{1.73}$$

Integrating Eq. (1.71) from $\xi - \epsilon$ to $\xi + \epsilon$, we find

$$\lim_{\epsilon \to 0} \left\{ \frac{dg}{dx}(\xi + \epsilon, \xi) - \frac{dg}{dx}(\xi - \epsilon, \xi) \right\} = 1. \tag{1.74}$$

This can be written as

$$\left[\!\!\left[\frac{dg}{dx}\right]\!\!\right]_{x=\xi} = 1. \tag{1.75}$$

The slope of g is discontinuous at $x = \xi$ (the quantity inside the double bracket denotes a **jump**), but g itself is continuous. We can enforce the continuity by choosing the constants A and B as

$$A = Cu_2(\xi), \quad B = Cu_1(\xi). \tag{1.76}$$

So far we have

$$g(x,\xi) = C \begin{cases} u_1(x)u_2(\xi), & x < \xi, \\ u_1(\xi)u_2(x), & x > \xi. \end{cases} \tag{1.77}$$

Using the jump condition (1.75), we find

$$C\{u_1(\xi)u_2'(\xi) - u_2(\xi)u_1'(\xi)\} = 1, \quad C\{\xi(-1) - (1-\xi)(1)\} = 1, \quad C = -1. \tag{1.78}$$

Finally,

$$g(x,\xi) = \begin{cases} x(\xi - 1), & x < \xi, \\ \xi(x - 1), & x > \xi, \end{cases} \tag{1.79}$$

which is the same as the Green's function we found by direct integration.

1.6 ADJOINT OPERATORS

Before we extend the idea of constructing Green's functions for arbitrary linear operators using the δ function as a forcing function, it is useful to extend the notion of symmetry in matrices to differential operators. With n-vectors x and y and an $n \times n$ matrix A, we find, if

$$y^T Ax = x^T B y, \tag{1.80}$$

then, as these are scalars, taking the transpose of one of them,

$$B = A^T, \tag{1.81}$$

and, if A is symmetric, $B = A$. Using the inner product notation, we have

$$\langle y, Ax \rangle = \langle x, By \rangle \Rightarrow \quad B = A^T. \tag{1.82}$$

The term "transpose" of a matrix operator translates into the "adjoint" of a differential operator. If we follow the terminology of matrix algebra, it is more reasonable to speak of the transpose of an operator. However, it is the convention to use the word "adjoint" for differential operators. Of course, the adjoint of a matrix is a totally different quantity. With L and L^* denoting two nth order linear differential operators and u and v functions in C^n, we call L^* the adjoint of L, if

$$\langle v, Lu \rangle = \langle u, L^*v \rangle, \tag{1.83}$$

with u and v satisfying appropriate homogeneous boundary conditions.

1.6.1 Example: Adjoint Operator

Find the adjoint operator and adjoint boundary conditions of the system,

$$Lu = x^2 u'' + u' + 2u, \quad u(1) = 0, \quad u'(2) + u(2) = 0. \tag{1.84}$$

For this system, we use integration-by-parts to apply the differential operator on v,

$$\langle v, Lu \rangle = \int_1^2 v[x^2 u'' + u' + 2u] \, dx$$

$$= \left\{ x^2 vu' - (x^2 v)'u + vu \right\} \Big|_1^2 + \int_1^2 u[(x^2 v)'' - v' + 2v] \, dx.$$

From the quantity inside the integral, we see

$$L^*v = (x^2 v)'' - v' + 2v = x^2 v'' + (4x - 1)v' + 4v,$$

$$L^* = x^2 \frac{d^2}{dx^2} + (4x - 1) \frac{d}{dx} + 4. \tag{1.85}$$

The quantity that has to be evaluated at the boundaries is called the bi-linear concomitant, $P(x)$, as it is linear in u and in v. The homogeneous boundary conditions have to be such that P should vanish at each boundary.

$$P(2) = 4v(2)u'(2) - 4v(2)u(2) - 4v'(2)u(2) + v(2)u(2)$$

$$= 4v(2)u'(2) - [3v(2) + 4v'(2)]u(2).$$

From the given condition, $u'(2) = -u(2)$, we have

$$P(2) = -[7v(2) + 4v'(2)]u(2).$$

Since $u(2)$ is arbitrary,

$$4v'(2) + 7v(2) = 0. \tag{1.86}$$

At the other boundary, we have

$$P(1) = v(1)u'(1) - 2v(1) + v'(1)u(1) + v(1)u(1)$$

$$= v(1)u'(1) - [v(1) - v'(1)]u(1).$$

Using, $u(1) = 0$, we get

$$v(1) = 0. \tag{1.87}$$

In general, L^* and the boundary conditions associated with it are different from L and its boundary conditions. When L^* is identical to L, we call L a self-adjoint operator. This case is analogous to a symmetric matrix operator.

1.7 GREEN'S FUNCTION AND ADJOINT GREEN'S FUNCTION

Let L and L^* be a linear operator and its adjoint with independent variable x. We assume there are associated homogeneous boundary conditions that render the bi-linear concomitant $P = 0$ at the boundaries. Consider

$$Lg(x,x_1) = \delta(x - x_1); \quad L^*g^*(x,x_2) = \delta(x - x_2), \tag{1.88}$$

where g^* is called the adjoint Green's function.

Now multiply the first equation by g^* and the second by g and form the inner products,

$$\langle g^*(x,x_2), Lg(x,x_1)\rangle - \langle g(x,x_1), L^*g^*(x,x_2)\rangle$$
$$= \langle g^*(x,x_2), \delta(x-x_1)\rangle - \langle g(x,x_1), \delta(x-x_2)\rangle. \qquad (1.89)$$

The left-hand side is zero by the definition of the adjoint system. After performing the integrations (remember, x is the independent variable), the right-hand side gives

$$g^*(x_1,x_2) = g(x_2,x_1) \quad \text{or} \quad g^*(\xi,x) = g(x,\xi). \qquad (1.90)$$

This shows the important symmetry between the Green's function and its adjoint. For the self-adjoint operator, $g^*(\xi,x) = g(x,\xi)$, and we have

$$g(x,\xi) = g(\xi,x). \qquad (1.91)$$

We saw this symmetry in the example where we had a self-adjoint operator, d^2/dx^2.

1.8 GREEN'S FUNCTION FOR L

Using the adjoint system in (a, b), again with x as the variable, for

$$Lu(x) = f(x), \quad L^*g^*(x,\xi) = \delta(x-\xi), \qquad (1.92)$$

by subtracting the inner products,

$$\langle g^*, Lu\rangle - \langle u, L^*g^*\rangle = \langle g^*, f(x)\rangle - \langle u, \delta(x-\xi)\rangle.$$

Again, the left-hand side is zero, and the right-hand side gives

$$u(\xi) = \int_a^b g^*(x,\xi)f(x)\,dx, \qquad (1.93)$$

which, after interchanging x and ξ, becomes

$$u(x) = \int_a^b g^*(\xi,x)f(\xi)\,d\xi. \qquad (1.94)$$

We can avoid g^* by using the symmetry between g and g^* and writing

$$u(x) = \int_a^b g(x,\xi)f(\xi)\,d\xi.$$

(1.95)

By applying the L-operator directly to this expression, we get

$$Lu = L\int_a^b g(x,\xi)f(\xi)\,d\xi = \int_a^b Lg(x,\xi)f(\xi)\,d\xi$$

$$= \int_a^b \delta(x-\xi)f(\xi)\,d\xi = f(x).$$

(1.96)

1.9 STURM-LIOUVILLE OPERATOR

A general self-adjoint second-order operator is the **Sturm-Liouville operator** L in the expression

$$Lu \equiv (pu')' + qu,$$

(1.97)

where $p(x)$ and $q(x)$ are given continuous functions with p being non-zero in (a, b). For various choices of p and q, $Lu = 0$ yields familiar functions such as the trigonometric functions ($p = 1, q = 1$), hyperbolic functions ($p = 1, q = -1$), Bessel functions ($p = x^2, q = n^2 - x^2$), Legendre functions ($p = 1 - x^2, q = -n(n+1)$), and so on. We assume certain homogeneous boundary conditions.

The Green's function for this operator has to satisfy

$$Lg(x,\xi) = \delta(x-\xi) \Rightarrow \quad Lg_1 = 0, \quad x < \xi, \quad Lg_2 = 0, \quad x > \xi,$$

(1.98)

with the same boundary conditions. Here we have used the notation

$$g = \begin{cases} g_1(x,\xi), & x < \xi, \\ g_2(x,\xi), & x > \xi. \end{cases}$$

(1.99)

Let u_1 and u_2 be two independent solutions of the homogeneous equation above, with u_1 satisfying the left boundary condition and u_2 satisfying the right boundary condition. Integrating

$$(pg')' + qg = \delta(x-\xi)$$

(1.100)

from $\xi - \epsilon$ to $\xi + \epsilon$, we find

$$(pg')\Big|_{\xi-\epsilon}^{\xi+\epsilon} + \int_{\xi-\epsilon}^{\xi+\epsilon} qg \, dx = 1. \qquad (1.101)$$

In Eq. (1.100) the δ function balances the first term, and qg is finite. As $\epsilon \to 0$, the second term in Eq. (1.101) goes to zero, and we obtain the jump condition

$$\left[\!\left[p(x)\frac{dg}{dx}(x,\xi) \right]\!\right]_{x=\xi} = 1,$$

$$\lim_{\epsilon\to 0}\left\{ g'(\xi+\epsilon,\xi) - g'(\xi-\epsilon,\xi) \right\} = \left[\!\left[\frac{dg}{dx}(x,\xi) \right]\!\right]_{x=\xi} = \frac{1}{p(\xi)}, \qquad (1.102)$$

where we have used the double bracket notation for the jump in slope.

Thus, the slope of the function g is discontinuous at ξ, but g itself is continuous at $x = \xi$. Now we set

$$g(x,\xi) = C \begin{cases} u_1(x)u_2(\xi), & x < \xi, \\ u_2(x)u_1(\xi), & x > \xi, \end{cases} \qquad (1.103)$$

which is continuous and symmetric. Using the jump condition, we find C as

$$C\left\{ u_2'(\xi)u_1(\xi) - u_2(\xi)u_1'(\xi) \right\} = \frac{1}{p(\xi)}. \qquad (1.104)$$

At first it appears C may be a function of ξ. But this is not the case. From

$$(pu_2')' + qu_2 = 0, \quad (pu_1')' + qu_1 = 0, \qquad (1.105)$$

we see the indefinite integral

$$\int^{x} [u_1(pu_2')' - u_2(pu_1')']\,dx = A, \qquad (1.106)$$

where A is a constant. If we integrate the first term by parts twice, we get

$$p(u_2'u_1 - u_2u_1') = A. \qquad (1.107)$$

This is called the **Abel identity**. It is worth noting that the Abel identity may be used as a first-order differential equation to find a second solution u_2 if we know only one solution u_1. The unknown C turns out be the reciprocal of the Abel constant in the form,

$$C = \frac{1}{A}. \tag{1.108}$$

We write the Green's function as

$$g(x,\xi) = \frac{1}{A} \begin{cases} u_1(x)u_2(\xi), & x < \xi, \\ u_2(x)u_1(\xi), & x > \xi. \end{cases} \tag{1.109}$$

The point ξ is known as the source point as it is the location of the delta function, and x is known as the observation point.

1.9.1 Method of Variable Constants

Having two independent solutions of the equation

$$(pu')' + qu = 0, \tag{1.110}$$

namely, u_1 and u_2, we can solve the nonhomogeneous equation

$$(pu')' + qu = f, \tag{1.111}$$

by the method of variable constants.

The method of variable constants assumes

$$u = A_1 u_1 + A_2 u_2, \tag{1.112}$$

where A_1 and A_2 are taken as functions of x. If we further stipulate that u_1 satisfies the left boundary condition, then A_2 can be set to zero at the left boundary. Similarly, assume u_2 satisfies the right boundary condition and A_1 is zero at the right boundary.

As there are two functions to be found and there is only one equation, Eq. (1.111), we impose the condition

$$u_1 A_1' + u_2 A_2' = 0. \tag{1.113}$$

Substituting Eq. (1.112) in the nonhomogeneous equation (1.111) and noting u_1 and u_2 satisfy the homogeneous equation, we get

$$pu_1'A_1' + pu_2'A_2' = f. \qquad (1.114)$$

Solutions of these two equations are

$$A_1'p(u_2'u_1 - u_2u_1') = -u_2f, \quad A_2'p(u_2'u_1 - u_2u_1') = u_1f. \qquad (1.115)$$

Using the Abel identity, these simplify to

$$A_1' = -\frac{1}{A}u_2f, \quad A_2' = \frac{1}{A}u_1f. \qquad (1.116)$$

Using the boundary conditions, $A_1(b) = 0$ and $A_2(a) = 0$, we integrate the preceding relations to get

$$A_1 = \frac{1}{A}\int_x^b u_2(\xi)f(\xi)\,d\xi, \quad A_2 = \frac{1}{A}\int_a^x u_1(\xi)f(\xi)\,d\xi. \qquad (1.117)$$

Now the solution, u, can be written as

$$u = \frac{1}{A}\left[\int_a^x u_1(\xi)u_2(x)f(\xi)\,d\xi + \int_x^b u_2(\xi)u_1(x)f(\xi)\,d\xi\right]. \qquad (1.118)$$

We extract the Green's function from this as

$$g(x,\xi) = \frac{1}{A}\left\{ \begin{array}{ll} u_1(x)u_2(\xi), & x < \xi, \\ u_2(x)u_1(\xi), & x > \xi, \end{array} \right. \qquad (1.119)$$

which is, of course, the same as the one we found using the δ function.

1.9.2 Example: Self-Adjoint Problem

Obtain the Green's function for the problem

$$(x^2u')' - \frac{3}{4}u = f(x), \qquad u(0) = 0, \quad u(1) = 0, \qquad (1.120)$$

and use it to solve the nonhomogeneous differential equation when $f(x) = \sqrt{x}$.

The given equation is in the self-adjoint form. Trying $u = x^\lambda$ as a solution for the homogeneous equation, we get

$$\lambda^2 + \lambda - 3/4 = 0, \quad \lambda = 1/2, \quad \lambda = -3/2. \tag{1.121}$$

The two solutions, $x^{1/2}$ and $x^{-3/2}$, can be used to get

$$u_1 = x^{1/2}, \quad u_2 = x^{1/2} - x^{-3/2}, \tag{1.122}$$

which satisfy the boundary conditions. In general, we may let

$$u_1 = A_1 x^{1/2} + B_1 x^{-3/2}, \quad u_2 = A_2 x^{1/2} + B_2 x^{-3/2}, \tag{1.123}$$

and obtain the constants to satisfy the boundary conditions.

The Green's function is written as

$$g(x,\xi) = C \begin{cases} x^{1/2}(\xi^{1/2} - \xi^{-3/2}), & x < \xi, \\ \xi^{1/2}(x^{1/2} - x^{-3/2}), & x > \xi. \end{cases} \tag{1.124}$$

Using the jump at $x = \xi$,

$$\left[\!\!\left[\frac{dg}{dx} \right]\!\!\right] = \frac{1}{p} = \frac{1}{\xi^2}, \tag{1.125}$$

we find $C = 1/2$. The solution of the nonhomogeneous problem when $f = \sqrt{x}$ is

$$u = \int_0^1 g(x,\xi)\sqrt{\xi}\, d\xi = \frac{1}{2}\left[\sqrt{x}\left(1 - \frac{1}{x^2}\right)\int_0^x \xi\, d\xi + \sqrt{x}\int_x^1 \left(\xi - \frac{1}{\xi}\right) d\xi \right],$$

$$u = \frac{1}{2}\sqrt{x}\log x. \tag{1.126}$$

1.9.3 Example: Non-Self-Adjoint Problem

Obtain the Green's function and its adjoint for the system,

$$Lu \equiv u'' - 2u' - 3u; \quad u(0) = 0, \quad u'(1) = 0. \tag{1.127}$$

From $\langle v, Lu\rangle - \langle u, L^*v\rangle = 0$ by integration by parts, we find

$$L^*v = v'' + 2v' - 3v; \quad v(0) = 0, \quad v'(1) + 2v(1) = 0. \tag{1.128}$$

Using solutions of the form, $e^{\lambda x}$, we have

$$Lu = 0, \quad \text{solutions:} \quad e^{-x}, e^{3x},$$

$$L^*v = 0, \quad \text{solutions:} \quad e^{x}, e^{-3x}.$$

We can combine these basic solutions to satisfy the boundary conditions, which gives

$$u_1 = e^{-x} - e^{3x}, \quad u_2 = e^{-x} + \frac{1}{3}e^{3x-4}, \tag{1.129}$$

$$v_1 = e^{x} - e^{-3x}, \quad v_2 = e^{x} + 3e^{4-3x}. \tag{1.130}$$

Integrating $Lu = \delta$ and $L^*v = \delta$, the jump conditions are

$$\left[\!\!\left[\frac{dg(x,\xi)}{dx}\right]\!\!\right]_{x=\xi} = 1, \quad \left[\!\!\left[\frac{dg^*(x,\xi)}{dx}\right]\!\!\right]_{x=\xi} = 1. \tag{1.131}$$

Next we assemble the Green's functions as

$$g(x,\xi) = C \begin{cases} (e^{-x} - e^{3x})(e^{-\xi} + \frac{1}{3}e^{3\xi-4}), & x < \xi, \\ (e^{-\xi} - e^{3\xi})(e^{-x} + \frac{1}{3}e^{3x-4}), & x > \xi, \end{cases} \tag{1.132}$$

$$g^*(x,\xi) = C^* \begin{cases} (e^{x} - e^{-3x})(e^{\xi} + 3e^{4-3\xi}), & x < \xi, \\ (e^{\xi} - e^{-3\xi})(e^{x} + 3e^{4-3x}), & x > \xi. \end{cases} \tag{1.133}$$

The values of C and C^* are found using the jump conditions, as

$$C\left\{(-e^{-\xi} + e^{3\xi-4})(e^{-\xi} - e^{3\xi}) - (-e^{-\xi} - 3e^{3\xi})\left(e^{-\xi} + \frac{1}{3}e^{3\xi-4}\right)\right\} = 1, \tag{1.134}$$

$$C^*\left\{\left(-3e^{4-3\xi} + \frac{1}{3}e^{\xi}\right)(e^{\xi} - e^{-3\xi}) - (e^{\xi} + 3e^{-3\xi})\left(e^{4-3\xi} + \frac{1}{3}e^{\xi}\right)\right\} = 1. \tag{1.135}$$

Simplifying these expressions, we get

$$C = \frac{3}{4}\frac{e^{-2\xi}}{3 + e^{-4}}, \quad C^* = -\frac{3}{4}\frac{e^{2\xi}}{1 + 3e^{4}}. \tag{1.136}$$

The two Green's functions are

$$g(x,\xi) = \frac{1}{4}\frac{1}{1+3e^4} \begin{cases} (e^{-x} - e^{3x})(3e^{4-3\xi} + e^{\xi}), & x < \xi, \\ (e^{-3\xi} - e^{\xi})(3e^{4-x} + e^{3x}), & x > \xi, \end{cases} \tag{1.137}$$

$$g^*(x,\xi) = \frac{1}{4}\frac{1}{1+3e^4} \begin{cases} (e^{-3x} - e^{x})(3e^{4-\xi} + e^{3\xi}), & x < \xi, \\ (e^{-\xi} - e^{3\xi})(3e^{4-3x} + e^{x}), & x > \xi. \end{cases} \tag{1.138}$$

We can observe the symmetry between g and g^*.

1.10 EIGENFUNCTIONS AND GREEN'S FUNCTION

We may use the eigenfunctions of the operators, L and L^*, with the associated homogeneous boundary conditions to solve the nonhomogeneous problem,

$$Lu = f. \tag{1.139}$$

Let u_n and v_n $(n = 1, 2, \ldots)$ be the eigenfunctions of L and L^*, respectively. Assume λ_n and μ_n are the sequences of eigenvalues associated with these eigenfunctions. That is

$$Lu_n = \lambda_n u_n, \quad L^* v_n = \mu_n v_n. \tag{1.140}$$

We assume that each of the sequence of eigenvalues are distinct and the eigenfunctions are complete. Then all of the v's cannot be orthogonal to a given u_n. Let us denote by v_n one of these v's that is not orthogonal to u_n. We now have

$$\lambda_n \langle v_n, u_n \rangle = \langle v_n, Lu_n \rangle = \langle u_n, L^* v_n \rangle = \mu_n \langle u_n, v_n \rangle, \tag{1.141}$$

which shows $\mu_n = \lambda_n$. The two operators, L and L^*, have the same eigenvalues,

$$Lu_n = \lambda_n u_n, \quad L^* v_m = \lambda_m v_m. \tag{1.142}$$

Forming inner products of the first equation with v_m and the second with u_n, we get

$$\langle v_m, Lu_n \rangle - \langle u_n, L^* v_m \rangle = (\lambda_n - \lambda_m)\langle v_m, u_n \rangle = 0. \tag{1.143}$$

This shows that for distinct eigenvalues, the eigenfunctions are bi-orthogonal,

$$\langle u_n, v_m \rangle = 0, \quad i \neq j. \tag{1.144}$$

Further, we normalize the eigenfunctions using the relation

$$\langle u_n, v_n \rangle = 1. \tag{1.145}$$

Getting back to the nonhomogeneous equation, $Lu = f$, we expand the unknown u and the given f as

$$u = \sum a_n u_n, \quad f = \sum b_n u_n, \quad b_n = \langle f, v_n \rangle, \tag{1.146}$$

where to get b_n, we used the bi-orthogonality of u_n and v_n:

$$Lu = L \sum a_n u_n = \sum a_n L u_n = \sum a_n \lambda_n u_n = \sum b_n u_n. \tag{1.147}$$

Since u_n are linearly independent, we get

$$a_n = \frac{1}{\lambda_n} b_n = \frac{1}{\lambda_n} \langle f, v_n \rangle. \tag{1.148}$$

The solution u is obtained as

$$u(x) = \sum \int_a^b \frac{u_n(x)v_n(\xi)}{\lambda_n} f(\xi) \, d\xi. \tag{1.149}$$

Comparing with the Green's function solution, we identify

$$g(x,\xi) = \sum \frac{u_n(x)v_n(\xi)}{\lambda_n}. \tag{1.150}$$

Further, using $g(x,\xi) = g^*(\xi,x)$, we have

$$g^*(x,\xi) = \sum \frac{v_n(x)u_n(\xi)}{\lambda_n}. \tag{1.151}$$

Of course, for a self-adjoint operator, $u_n = v_n$ and $g = g^*$.

1.10.1 Example: Eigenfunctions

Consider the equation

$$\frac{d^2u}{dx^2} = f, \quad u(0) = 0, \quad u(1) = 0. \tag{1.152}$$

The eigenvalue problem for this self-adjoint system is

$$\frac{d^2u_n}{dx^2} = \lambda_n u_n, \quad u_n(0) = 0, \quad u_n(1) = 0; \quad n = 1, 2, \ldots. \tag{1.153}$$

Let $\lambda = -\mu_n^2$. The solution u_n is found as

$$u_n = A_n \cos(\mu_n x) + B_n \sin(\mu_n x). \tag{1.154}$$

The boundary conditions give

$$A_n = 0, \quad B_n \sin(\mu_n) = 0. \tag{1.155}$$

For nontrivial solutions, $B_n \neq 0$, and we must have

$$\sin(\mu_n) = 0, \quad \mu_n = n\pi, \quad \lambda_n = -n^2\pi^2. \tag{1.156}$$

We choose $B_n = \sqrt{2}$, so that $\|u_n\| = 1$. The Green's function has the eigenfunction expansion

$$g(x, \xi) = -\sum_{n=1}^{\infty} \frac{2\sin(n\pi x)\sin(n\pi \xi)}{n^2\pi^2}. \tag{1.157}$$

This is nothing but a Fourier series representation of the the Green's function, (1.79).

1.11 HIGHER-DIMENSIONAL OPERATORS

In many applications, we encounter the generalized Sturm-Liouville equation

$$Lu = \frac{\partial}{\partial x}\left(p\frac{\partial u}{\partial x}\right) + \frac{\partial}{\partial y}\left(p\frac{\partial u}{\partial y}\right) + \frac{\partial}{\partial z}\left(p\frac{\partial u}{\partial z}\right) + qu = f(x,y,z), \tag{1.158}$$

in a three dimensional (3D) domain Ω with homogeneous conditions on the boundary $\partial\Omega$. Here, p and q are functions of x, y, and z. We will also consider the two-dimensional version where the z-dependence is absent. Defining the gradient operator

$$\nabla = i\frac{\partial}{\partial x} + j\frac{\partial}{\partial y} + k\frac{\partial}{\partial z},$$

(1.159)

where i, j, and k are the cartesian unit vectors, we can write the Sturm-Liouville equation as

$$\nabla \cdot (p\nabla u) + qu = f.$$

(1.160)

Let

$$n = in_x + jn_y + kn_z$$

(1.161)

be the outward normal to the boundary surface $\partial\Omega$.

The inner product is now defined as the volume integral

$$\langle u,v \rangle = \int_\Omega uv d\Omega.$$

(1.162)

As before, let the Green's function, $g(x,\xi)$, satisfy

$$Lg = \delta(x - \xi).$$

(1.163)

Then

$$\langle g, Lu \rangle - \langle u, Lg \rangle = 0 = \langle g, f \rangle - \langle u, \delta \rangle,$$

(1.164)

provided g satisfies the homogeneous boundary conditions on $\partial\Omega$. We now have

$$u(x) = \int_\Omega g(x,\xi)f(\xi)d\Omega.$$

(1.165)

Formally, this shows the Green's function representation of the solution of any self-adjoint partial differential equation in an n-dimensional domain. The extension of this representation for any non-self-adjoint operator is similar to what was done in the case of the one-dimensional equations.

To see the behavior of the Green's function in the neighborhood of the source point $\boldsymbol{\xi}$, we consider a small spherical volume, V, of radius $r = \epsilon$ centered at $\boldsymbol{\xi}$ and integrate the equation, $Lu = \delta$, over this volume. Using the fact that the integral of the delta function is unity, we see

$$\int_V [\nabla \cdot (p\nabla g) + qg]\, dV = 1. \tag{1.166}$$

Using the Gauss theorem, the first term of the integral can be converted to a surface integral,

$$\int_S \boldsymbol{n} \cdot p\nabla g\, dS + \int_V qg\, dV = 1,$$

$$\int_S p\frac{\partial g}{\partial r}\, dS + \int_V qg\, dV = 1, \tag{1.167}$$

where we have used the directional derivative

$$\boldsymbol{n} \cdot p\nabla g = p\frac{\partial g}{\partial r}, \tag{1.168}$$

with r representing a coordinate normal to the surface (i.e., $r = |\boldsymbol{x} - \boldsymbol{\xi}|$). As $\epsilon \to 0$, we observe that the volume integral of qg is small compared to the surface integral, and the Green's function is spherically symmetrical as the boundaries are far away compared to the scale of ϵ. From this the behavior of the Green's function for small values of r is of the form,

$$4\pi pr^2\frac{dg}{dr} \sim 1, \quad \frac{dg}{dr} \sim \frac{1}{4\pi pr^2}. \tag{1.169}$$

For an infinite domain the boundary effects are absent, and, in addition, if $q = 0$, the preceding result can be written as

$$4\pi pr^2\frac{dg}{dr} = 1, \quad \frac{dg}{dr} = \frac{1}{4\pi pr^2}. \tag{1.170}$$

Further, if $p = 1$, the exact Green's function for an infinite domain becomes

$$g_\infty = -\frac{1}{4\pi r}, \quad r = \{(x - \xi)^2 + (y - \eta)^2 + (z - \zeta)^2\}^{1/2}. \tag{1.171}$$

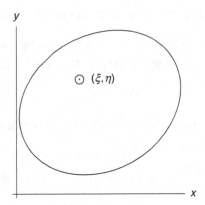

Figure 1.8. Two-dimensional domain.

With $p = 1$ and $q = 0$, the Sturm-Liouville equation becomes the Poisson equation

$$\nabla^2 u = f. \tag{1.172}$$

We could apply the above integration using the Gauss theorem for the two-dimensional (2D) Sturm-Liouville equation (see Fig. 1.8). This results in

$$2\pi pr \frac{dg}{dr} \sim 1, \quad \frac{dg}{dr} \sim \frac{1}{2\pi pr}. \tag{1.173}$$

Again, when $p = 1$, $q = 0$ we have the Poisson equation,

$$\nabla^2 u = f, \tag{1.174}$$

with the exact Green's function for the infinite domain,

$$g_\infty = \frac{1}{2\pi} \log r, \quad r = \{(x - \xi)^2 + (y - \eta)^2\}^{1/2}. \tag{1.175}$$

Now we have exact Green's functions for the Laplace operators in 2D and 3D infinite spaces. To obtain the solution u in terms of g_∞, we need to compute the integrals of f multiplied by g over the whole space. For these integrals to exist, certain conditions on the decay of f at infinity are required. Of course, in bounded domains, g_∞ does not satisfy the boundary conditions, and we have to resort to other methods.

1.11.1 Example: Steady-State Heat Conduction in a Plate

Consider an infinite plate under steady-state temperature distribution with a heat source distribution, $q(x,y)$. The temperature, T, satisfies

$$\nabla^2 T = -\frac{q}{k}, \qquad (1.176)$$

where k is the conductivity. Using the two-dimensional Green's function, the solutions is written as

$$T(x,y) = -\frac{1}{4\pi k} \int_{-\infty}^{\infty} \int_{-\infty}^{\infty} q(\xi,\eta) \log[(x-\xi)^2 + (y-\eta)^2] d\xi\, d\eta. \quad (1.177)$$

Usually, the source is limited to a finite area, and the limits of the above integral will have finite values. If the heat source has a circular boundary, polar coordinates may be more convenient.

1.11.2 Example: Poisson's Equation in a Rectangle

To solve the Poisson's equation

$$\nabla^2 u = f(x,y); \quad -a < x < a, \quad -b < y < b, \qquad (1.178)$$

with $u = 0$ on the rectangular boundary, we need the Green's function satisfying the same boundary condition. Here, we illustrate the eigenfunction representation of the Green's function for this purpose. Let u_{mn} (where m and n will take integer values) represent an eigenfunction of the Laplace equation with $u_{mn} = 0$ on the boundary. We have

$$\frac{\partial u_{mn}}{\partial x^2} + \frac{\partial u_{mn}}{\partial y^2} = \lambda_{mn} u_{mn}. \qquad (1.179)$$

Using separation of variable, we represent u_{mn} as

$$u_{mn}(x,y) = X_m(x) Y_n(y). \qquad (1.180)$$

Substituting this in the Laplace equation and dividing everything by $X_m Y_n$, we get

$$\frac{X_m''}{X_m} + \frac{Y_n''}{Y_n} = \lambda_{mn}. \qquad (1.181)$$

Let

$$\frac{X_m''}{X_m} = -\mu_m^2, \quad \frac{Y_n''}{Y_n} = -v_n^2, \quad \lambda_{mn} = -(\mu_m^2 + v_n^2). \tag{1.182}$$

Solutions of these equations with $X_m(\pm a) = 0$ and $Y_n(\pm b) = 0$ are

$$X_m = \sin m\pi x/a, \quad Y_n = \sin n\pi y/b; \quad \mu_m = m\pi/a, \quad v_n = n\pi/b. \tag{1.183}$$

By integrating these functions over their respective intervals, we can make their norms unity if we scale these as

$$X_m = \frac{1}{\sqrt{a}}\sin\frac{m\pi x}{a}, \quad Y_n = \frac{1}{\sqrt{b}}\sin\frac{n\pi y}{b}. \tag{1.184}$$

The two-dimensional eigenfunctions and eigenvalues are

$$u_{mn}(x,y) = \frac{1}{\sqrt{ab}}\sin\frac{m\pi x}{a}\sin\frac{n\pi y}{b}, \quad \lambda_{mn} = -\pi^2\left(\frac{m^2}{a^2} + \frac{n^2}{b^2}\right). \tag{1.185}$$

The Green's function becomes

$$g(x,y,\xi,\eta) = \sum_{m=1}^{\infty}\sum_{n=1}^{\infty}\frac{u_{mn}(x,y)u_{mn}(\xi,\eta)}{\lambda_{mn}}. \tag{1.186}$$

The solution of the nonhomogeneous equation using this Green's function is identical to the one we could obtain using a Fourier series approach.

1.11.3 Steady-State Waves and the Helmholtz Equation

The wave equation for a field variable, Ψ, propagating with a speed c is given by

$$\nabla^2\Psi = \frac{1}{c^2}\frac{\partial^2\Psi}{\partial t^2}. \tag{1.187}$$

If the waves are harmonic, we can separate the time dependence and space dependence in the form

$$\Psi = \psi(x,y,z)e^{-i\Omega t}, \tag{1.188}$$

where Ω is the angular frequency, and ψ satisfies

$$\nabla^2\psi + k^2\psi = 0, \quad k = \Omega/c, \tag{1.189}$$

with k representing the wave number. This equation is called the **Helmholtz equation**. From our discussion of the Sturm-Liouville problem, now we have $p = 1$ and $q = k^2$, and the singular behavior of the Green's function is unaffected. By direct substitution, it can be verified that

$$g = -\frac{e^{ikr}}{4\pi r} \tag{1.190}$$

is the required Green's function for the outgoing waves (the exponent, $kr - \Omega t$, in Ψ can be kept constant if r and t both increase by Δr and $\Delta r/c$, respectively).

For the 2D case,

$$g = -\frac{i}{4}H_0^{(1)}(kr), \tag{1.191}$$

which is one of the Hankel functions

$$H_0^{(1)} = J_0 + iY_0, \quad H_0^{(2)} = J_0 - iY_0, \tag{1.192}$$

with J_0 and Y_0 being the Bessel and Neumann functions, respectively. The Hankel functions have logarithmic behavior through the Neumann function Y_0, as $r \to 0$. We use $H_0^{(1)}$ for outgoing waves and $H_0^{(2)}$ for incoming waves.

1.12 METHOD OF IMAGES

We could use the Green's function, g_∞, for the 2D infinite domain, obtained here to solve the Laplace equation in other domains, in two ways: one is the method of images, which is useful if the new domain can be obtained by symmetrically folding the full infinite domain and the other is through conformal mapping.

As shown in Fig. 1.9, to construct the Green's function for the semi-infinite domain, $x > 0$, with $g = 0$ at $x = 0$, we introduce the usual source

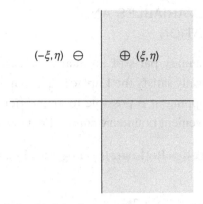

Figure 1.9. Semi-infinite domain with a source and its image.

at (ξ,η) and an image sink at $(-\xi,\eta)$. The sink is outside the domain, but the combined effect is to have

$$g = \frac{1}{4\pi}\mathrm{Log}\frac{(x-\xi)^2+(y-\eta)^2}{(x+\xi)^2+(y-\eta)^2},\qquad (1.193)$$

equals to zero on $x=0$. Here, we have extended g_∞ into $x<0$ in an odd fashion. If we needed g with normal derivative zero on $x=0$, we have to extend g_∞ in an even fashion, using two sources.

To obtain the Green's function for a quarter plane we need four sources (two of them may be sinks depending on the boundary conditions). Similarly, for a 45° wedge, we use eight sources at the points, (ξ,η), (η,ξ) and at the images of these two points under reflection with respect to the x- and y-axes.

We can even obtain the Green's function for a rectangle with one corner at $(0,0)$ and the diagonally opposite corner at (a,b) by repeatedly reflecting the original source at (ξ,η) about (a) the $x=0$ line, (b) $x=a$ line, (c) $y=0$ line, and (d) $y=b$ line. This gives a doubly infinite system of sources.

For the 3D case, if the new domain can be obtained by symmetrically sectioning the full infinite domain, we can construct the Green's function by the method of images.

1.13 COMPLEX VARIABLES AND THE LAPLACE EQUATION

The real and imaginary parts of an analytic function of a complex variable automatically satisfy the Laplace equation in 2D. Also, using the conformal mapping, it is possible to map different domains into domains with convenient boundary curves. First, let us note that

$$2\pi g_\infty = \mathrm{Re}[\mathrm{Log}(z)] = \mathrm{Log}\,|z| = \mathrm{Log}\,r, \qquad (1.194)$$

satisfies

$$\nabla^2 g_\infty = \delta(x,y). \qquad (1.195)$$

We can move the source to $\zeta = \xi + i\eta$ by defining

$$2\pi g_\infty = \mathrm{Log}\,|z - \zeta|. \qquad (1.196)$$

For a semi-infinite domain, $0 < y$, with $g = 0$ on the real axis $y = 0$, using the method of images, we introduce a sink at $\xi - i\eta = \bar{\zeta}$ and write

$$2\pi g = \mathrm{Log}\left|\frac{z - \zeta}{z - \bar{\zeta}}\right|. \qquad (1.197)$$

We can map (see Fig. 1.10) the upper half plane onto the interior of a unit circle, with the line $y = 0$ becoming the circular boundary, using

$$w = \frac{i - z}{i + z}, \quad \text{or} \quad z = \frac{1}{i}\frac{w - 1}{w + 1}. \qquad (1.198)$$

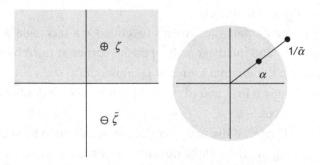

Figure 1.10. Mapping the upper half plane to the interior of a circle.

Let α be the image of the source point ζ. Then

$$\zeta = \frac{1}{i}\frac{\alpha-1}{\alpha+1}. \tag{1.199}$$

Using these in the expression for g, after some simplification, we get

$$2\pi g = \text{Log}\left|\frac{w-\alpha}{1-\bar{\alpha}w}\right|. \tag{1.200}$$

Now we have the Green's function for a unit circle with g being zero on the boundary.

Another way to view this Green's function is to consider

$$2\pi g = \text{Log}\,|z|, \tag{1.201}$$

as the solution for the Laplace equation with the right type of singularity at $z=0$. The transformation

$$z = e^{i\phi}\frac{w-\alpha}{1-\bar{\alpha}w} \tag{1.202}$$

moves the origin to α but retains the circular boundary. The angle ϕ gives a rigidbody rotation.

Using the notation

$$w = re^{i\theta}, \quad \alpha = \rho e^{i\phi}, \tag{1.203}$$

we can write

$$g(r,\theta,\rho,\phi) = \frac{1}{4\pi}\text{Log}\frac{r^2 - 2r\rho\cos(\phi-\theta)+\rho^2}{1-2r\rho\cos(\phi-\theta)+r^2\rho^2}, \tag{1.204}$$

which solves the Laplace equation on a unit circle. In this form, it is easy to see that g is indeed zero when $r=1$.

Conformal mapping can be used to map domains onto a unit circle and the Green's function, Eq. (1.204), can be used to solve the Poisson equation. In particular, the **Schwartz-Christoffel transform** maps polygons onto the upper half plane.

1.13.1 Nonhomogeneous Boundary Conditions

Consider the Poisson equation

$$\nabla^2 u = f, \quad (x,y) \in \Omega, \tag{1.205}$$

with the boundary condition

$$u = h, \quad (x,y) \in \partial\Omega. \tag{1.206}$$

Let g satisfy

$$\nabla^2 g = \delta(x - \xi, y - \eta), \quad g = 0 \quad \text{on} \quad (x,y) \in \partial\Omega. \tag{1.207}$$

The inner products give

$$\langle g, \nabla^2 u \rangle - \langle u, \nabla^2 g \rangle = \oint \left[g \frac{\partial u}{\partial n} - u \frac{\partial g}{\partial n} \right] ds. \tag{1.208}$$

As $g = 0$ on the boundary, the first term on the right is zero, and we find

$$u(\xi, \eta) = \int_\Omega g(x,y,\xi,\eta) f(x,y) \, dx \, dy + \oint h \frac{\partial g}{\partial n} \, ds. \tag{1.209}$$

As long as $g = 0$ on the boundary, we can incorporate nonhomogeneous boundary conditions without any complications.

1.13.2 Example: Laplace Equation in a Semi-infinite Region

To solve

$$\nabla^2 u = 0, \quad -\infty < x < \infty, \quad 0 < y < \infty, \tag{1.210}$$

with $u = f(x)$ on the boundary $y = 0$, we use

$$g = \frac{1}{2\pi} \text{Log} \left| \frac{z - \zeta}{z - \bar{\zeta}} \right|. \tag{1.211}$$

Assuming u tends to zero at infinity, Eq. (1.209) becomes

$$u(\xi, \eta) = \frac{1}{2\pi i} \int_{-\infty}^{\infty} \left[\frac{1}{z - \zeta} - \frac{1}{z - \bar{\zeta}} \right] f(x) \, dx, \quad y = 0. \tag{1.212}$$

Before we simplify this, it is worth noting that the right-hand side may be considered as two Cauchy integrals evaluated over a closed curve consisting of the real axis and an infinite semicircle in the upper half plane: one with a pole at $z = \zeta$ inside this contour and the other with a pole at $z = \bar{\zeta}$ outside the contour. The integrals over the infinite semicircle are assumed to go to zero.

The simplified version is usually written as

$$u(x,y) = \frac{y}{\pi} \int_{-\infty}^{\infty} \frac{f(\xi)d\xi}{(x-\xi)^2 + y^2}. \tag{1.213}$$

1.13.3 Example: Laplace Equation in a Unit Circle

To solve the Laplace equation in a unit circle with $u = h(\theta)$ on the boundary, we use the Green's function of Eq. (1.204) to find

$$\frac{\partial g}{\partial n} = \frac{\partial g}{\partial \rho}$$

$$= \frac{1}{2\pi} \frac{1-r^2}{1-2r\cos(\phi-\theta)+r^2}. \tag{1.214}$$

The solution is given by

$$u(r,\theta) = \frac{1-r^2}{2\pi} \int_0^{2\pi} \frac{h(\phi)\,d\phi}{1-2r\cos(\phi-\theta)+r^2}. \tag{1.215}$$

This is known as the Poisson representation of the solution of the Laplace equation. This can also be interpreted in terms of the Cauchy integrals.

1.14 GENERALIZED GREEN'S FUNCTION

There are problems, such as

$$u'' = f(x), \quad u'(0) = 0, \quad u'(1) = 0, \tag{1.216}$$

where a solution of the homogeneous equation, $u'' = 0$, namely,

$$U(x) = 1, \tag{1.217}$$

satisfies both the boundary conditions. In this case, our récipé for constructing the Green's function, which calls for a left and a right solution with a jump in slope at the source point, fails. It can be seen that

$$\langle U, u'' \rangle - \langle u, U'' \rangle = 0 \qquad (1.218)$$

implies the existence condition

$$\langle U, f \rangle = 0. \qquad (1.219)$$

In other words, the nonhomogeneous problem does not have a solution for all given functions f, but only for a restricted class of functions orthogonal to U. Further, with any particular solution, u_p, we get for the equation, we can add a constant times U to get the nonunique general solution

$$u = u_p + AU(x). \qquad (1.220)$$

To make u_p unique, we choose it to be orthogonal to U.

For the Sturm-Liouville operator, L, we want to solve

$$Lu = f, \qquad (1.221)$$

with certain homogeneous conditions at $x = a$ and $x = b$. We have a homogeneous solution, U, satisfying

$$LU = 0, \qquad (1.222)$$

with the same homogeneous conditions. Note that $U(x)$ is an eigenfunction corresponding to an eigenvalue of zero. We also stipulate the existence condition,

$$\langle U, f \rangle = 0. \qquad (1.223)$$

The differential equation for g has to be different from the one we used before. Let

$$Lg = \delta(x - \xi) + h(x), \qquad (1.224)$$

where h has to be found. Using the existence condition on this equation, we see

$$\langle U, \delta(x-\xi) + h(x) \rangle = 0, \quad U(\xi) + \int_a^b h(x)U(x)\,dx = 0. \qquad (1.225)$$

This implies that the unknown $h(x)$ has the form

$$h(x) = U(\xi)w(x), \qquad (1.226)$$

where $w(x)$ has to be found, and

$$\int_a^b w(x)U(x)\,dx = -1. \qquad (1.227)$$

Forming the difference of the inner products, we get

$$\langle g, Lu \rangle - \langle u, Lg \rangle = 0 = \langle g, f \rangle - \langle u, \delta + U(\xi)w \rangle. \qquad (1.228)$$

From this we get

$$u(\xi) + U(\xi)\langle u, w \rangle = \langle g, f \rangle. \qquad (1.229)$$

By noting u can be made orthogonal to U, we let

$$w = CU(x). \qquad (1.230)$$

This choice in conjunction with the existence condition, (1.227), on w gives

$$C = -1/\|U\|^2. \qquad (1.231)$$

If we use a normalized solution U, with $\|U\| = 1$, then $C = -1$. Equation (1.229) now has the required form

$$u(x) = \int_a^b g(x,\xi)f(\xi)\,d\xi, \qquad (1.232)$$

where g satisfies

$$Lg = \delta(x-\xi) - U(x)U(\xi), \quad \|U\| = 1, \qquad (1.233)$$

with all three functions, u, g, and U satisfying the same homogeneous boundary conditions. We note that a multiple of U can be added to u to get the general solution.

1.14.1 Examples: Generalized Green's Functions

(a) Consider the equation

$$u'' = f(x), \quad u'(0) = u'(1) = 0. \tag{1.234}$$

The function

$$U(x) = 1 \tag{1.235}$$

satisfies both boundary conditions and the homogeneous equation. This calls for the generalized Green's function, which satisfies

$$g'' = \delta(x - \xi) - U(x)U(\xi), \quad \text{i.e.,} \quad g'' = \delta(x - \xi) - 1. \tag{1.236}$$

Considering g in two parts: g_1 and g_2 for $x < \xi$ and $x > \xi$, we find

$$g_1' = -x, \quad g_2' = 1 - x. \tag{1.237}$$

Integrating again

$$g_1 = -\frac{x^2}{2} + D_1, \quad g_2 = -\frac{(x-1)^2}{2} + D_2. \tag{1.238}$$

The continuity at $x = \xi$ is satisfied by choosing

$$D_1 = D - \frac{(\xi - 1)^2}{2}, \quad D_2 = D - \frac{\xi^2}{2}. \tag{1.239}$$

Letting $\langle U, g \rangle = 0$, we find $D = 1/6$. Finally,

$$g(x, \xi) = \frac{1}{6} - \frac{1}{2} \begin{cases} x^2 + (\xi - 1)^2, & x < \xi, \\ \xi^2 + (x - 1)^2, & x > \xi. \end{cases} \tag{1.240}$$

(b) Next, consider

$$u'' + u = f(x), \quad u(0) = 0, \quad u(\pi) = 0. \tag{1.241}$$

The normalized solution of the homogeneous equation, which satisfies the boundary conditions, is

$$U(x) = \sqrt{\frac{2}{\pi}} \sin x. \tag{1.242}$$

The generalized Green's function satisfies

$$g'' + g = \delta(x - \xi) - \frac{2}{\pi} \sin x \sin \xi, \qquad (1.243)$$

with the same homogeneous boundary conditions.

Considering g in two parts,

$$g_1 = \frac{1}{\pi}[x \cos x \sin \xi + D_1 \sin x], \quad g_2 = \frac{1}{\pi}[(x - \pi) \cos x \sin \xi + D_2 \sin x]. \qquad (1.244)$$

Continuity of g can be satisfied by taking

$$D_1 = D + (\xi - \pi) \cos \xi, \quad D_2 = D + \xi \cos \xi. \qquad (1.245)$$

At this stage we have

$$\pi g(x, \xi) = D \sin x + \begin{cases} x \cos x \sin \xi + (\xi - \pi) \cos \xi \sin x, & x < \xi, \\ (x - \pi) \cos x \sin \xi + \xi \cos \xi \sin x, & x > \xi. \end{cases} \qquad (1.246)$$

After completing the integral, $\langle g, U \rangle = 0$, we find

$$D = -\frac{1}{2} \sin \xi, \qquad (1.247)$$

and

$$g(x, \xi) = -\frac{1}{2\pi} \sin x \sin \xi + \frac{1}{\pi} \begin{cases} x \cos x \sin \xi + (\xi - \pi) \cos \xi \sin x, & x < \xi, \\ (x - \pi) \cos x \sin \xi + \xi \cos \xi \sin x, & x > \xi. \end{cases} \qquad (1.248)$$

The jump condition is automatically satisfied.

1.14.2 A Récipé for Generalized Green's Function

From the examples given earlier, we write down a récipé for constructing the generalized Green's functions for the Sturm-Liouville operator when separate left and right boundary conditions are given, as follows:

1.1 Obtain $U(x)$, which satisfies the homogeneous equation and both boundary conditions. Normalize U.

1.2 Find $u_1(x)$ and $u_2(x)$, which satisfy the equation

$$Lu = -U(x), \qquad (1.249)$$

with u_1 satisfying the left boundary condition and u_2 satisfying the right boundary condition.

1.3 Let

$$g_1 = DU(x)U(\xi) + u_1(x)U(\xi) + U(x)u_2(\xi),$$
$$g_2 = DU(x)U(\xi) + u_2(x)U(\xi) + U(x)u_1(\xi). \qquad (1.250)$$

1.4 Find D using $\langle g, U \rangle = 0$.

For the Sturm-Liouville operator, using

$$Lu_1 = -U, \quad Lu_2 = -U, \qquad (1.251)$$

and performing

$$\int_a^\xi (ULu_1 - u_1LU)\,dx + \int_\xi^b (ULu_2 - u_2LU)\,dx = -\int_a^b U^2 dx = -1, \qquad (1.252)$$

the left-hand side gives

$$p(\xi)\{[u_1'(\xi) - u_2'(\xi)]U(\xi) - [u_1(\xi) - u_2(\xi)]U'(\xi)\} = -1. \qquad (1.253)$$

Comparing this with $g_2' - g_1'$, we see the generalized Green's function has a jump of $1/p$ at $x = \xi$.

1.15 NON-SELF-ADJOINT OPERATOR

The procedure for finding the generalized Green's function for the Sturm-Liouville operator can be extended to a general operator L as follows.

Consider L and its adjoint L^* with two sets of homogeneous boundary conditions (say, H and H^*) that make the bi-linear concomitant zero. We seek a generalized Green's function when there exists a function $U(x)$ such that

$$LU = 0, \tag{1.254}$$

and U satisfies the boundary conditions H.

In this case, the equation

$$L^*U^* = 0 \tag{1.255}$$

also has a solution U^*, which satisfies the boundary conditions H^*. We come to this conclusion through two observations: (a) L has an eigenfunction U with eigenvalue zero, and L^* must also have a zero eigenvalue; (b) If $L^*g^* = \delta$ allows a left and right solutions, we can use g^* to solve our original problem $Lu = f$ uniquely, which we know is false.

As in the case of the Sturm-Liouville problem, we begin with

$$Lu = f, \quad L^*U^* = 0. \tag{1.256}$$

From this we get the existence condition

$$\langle U^*, f \rangle = 0. \tag{1.257}$$

The solution u will not be unique as additive terms, which are multiples of $U(x)$, are allowed. We may choose a particular solution orthogonal to U^*, that is,

$$\langle U^*, u \rangle = 0. \tag{1.258}$$

Note that in non-self-adjoint systems, we go by bi-orthogonality. We assume the generalized Green's functions satisfy

$$Lg = \delta(x - \xi) + h(x), \quad L^*g^* = \delta(x - \xi) + h^*(x), \tag{1.259}$$

where h and h^* have to be found. From

$$\langle U, L^*g^* \rangle - \langle g^*, LU \rangle = 0 = \langle U, \delta(x - \xi) + h^* \rangle \tag{1.260}$$

we get

$$U(\xi) + \int_a^b U h^* dx = 0. \tag{1.261}$$

This allows us to take

$$h^*(x) = U(\xi) w^*(x), \qquad \int_a^b w^*(x) U(x) dx = -1. \tag{1.262}$$

Similarly

$$h(x) = U^*(\xi) w(x), \qquad \int_a^b w(x) U^*(x) dx = -1. \tag{1.263}$$

From

$$\langle g^*, Lu \rangle - \langle u, L^* g^* \rangle = 0 = \langle g^*, f \rangle - \langle u, \delta + U(\xi) w^* \rangle, \tag{1.264}$$

we get

$$u(\xi) + U(\xi) \int_a^b u(x) w^*(x) dx = \int_a^b g^*(x, \xi) f(x) dx. \tag{1.265}$$

Using Eq. (1.258), we select

$$w^*(x) = -U^*(x), \quad w(x) = -U(x). \tag{1.266}$$

Here, the negative signs are obtained from Eqs. (1.262) and (1.263), with the normalization

$$\langle U, U^* \rangle = 1. \tag{1.267}$$

Thus, the generalized Green's functions satisfy

$$Lg = \delta(x - \xi) - U^*(\xi) U(x), \quad L^* g^* = \delta(x - \xi) - U(\xi) U^*(x). \tag{1.268}$$

Using $\xi = \xi_1$ in the first equation and $\xi = \xi_2$ in the second equation,

$$Lg = \delta(x - \xi_1) - U^*(\xi_1) U(x), \quad L^* g^* = \delta(x - \xi_2) - U(\xi_2) U^*(x), \tag{1.269}$$

we find

$$g^*(\xi_1, \xi_2) = g(\xi_2, \xi_1), \tag{1.270}$$

where we have used the existence conditions

$$\langle g^*, U \rangle = 0 = \langle g, U^* \rangle. \tag{1.271}$$

From the symmetry of g and g^*, Eq. (1.265) can be written as

$$u(x) = \int_a^b g(x,\xi) f(\xi) d\xi + A U(x), \tag{1.272}$$

where we have added a non-unique term with an arbitrary constant A, to cast u in the general form.

When there are more than one eigenfunction corresponding to the zero eigenvalue, these eigenfunctions must be included in the construction of the Green's function, in a manner similar to what has been done here with one eigenfunction.

1.16 MORE ON GREEN'S FUNCTIONS

In Chapter 2, we mention numerical methods based on integral equations for using Green's functions developed for infinite domains for problems defined in finite domains. In Chapters 3 and 4, using the Fourier and Laplace transform methods, additional Green's functions are developed. These include the Green's functions of heat conduction and wave propagation problems. There is an extensive literature concerning the use of Green's functions in quantum mechanics, and the famous Feynman diagrams deal with perturbation expansions of Green's functions.

SUGGESTED READING

Abramowitz, M., and Stegun, I. (1965). *Handbook of Mathematical Functions* (National Bureau of Standards), Dover.

Beck, J., Cole, K., Haji-Sheikh, A., and Litkouhi, B. (1992). *Heat Conduction Using Green's Functions*, Hemisphere.

Courant, R., and Hilbert, D. (1953). *Methods of Mathematical Physics*, Vol. I, Interscience.

Hildebrand, F. B. (1992). *Methods of Applied Mathematics*, Dover.

Morse, P. M., and Feshbach, H. (1953). *Methods of Theoretical Physics*, Vol. I, McGraw-Hill.
Stakgold, I. (1968). *Boundary Value Problems of Mathematical Physics*, Vol. 1 and 2, Mcmillan.

EXERCISES

1.1 The deflection of a beam is governed by the equation

$$EI\frac{d^4v}{dx^4} = -p(x),$$

where EI is the bending stiffness and $p(x)$ is the distributed loading on the beam. If the beam has a length ℓ, and at both the ends the deflection and slope are zero, obtain expressions for the deflection by direct integration, using the Macaulay brackets when necessary, if

(a) $p(x) = p_0$,
(b) $p(x) = P_0\delta(x - \xi)$,
(c) $p(x) = M_0\delta'(x - \xi)$.

Obtain the Green's function for the deflection equation from the preceding calculations.

1.2 Solve the preceding problem when the beam is simply supported. That is,

$$v(0) = v(\ell) = 0, \quad v''(0) = v''(\ell) = 0.$$

1.3 Obtain the derivative of the function

$$g(x) = |f(x)|,$$

in $a < x < b$, assuming $f(x)$ has a simple zero at the point c inside the interval (a,b). Use the Signum function and/or delta function to express the result.

1.4 Assuming a function $f(x)$ has simple zeros at $x_i, i = 1,2,\ldots,n$, find an expression for

$$\delta(f(x)).$$

1.5 Convert the equation

$$Lu = \frac{d^2u}{dx^2} + x^2\frac{du}{dx} + 2u = f$$

into the Sturm-Liouville form.

1.6 Find the adjoint system for

$$xu'' + u' + u = 0, \quad u(1) = 0, \quad u(2) + u'(2) = 0.$$

1.7 Solve the differential system

$$u'' + u' - 2u = x^2, \quad u(0) = 0, \quad u'(1) = 0.$$

1.8 Solve the differential system

$$(x^2u')' - n(n+1)u = 0, \quad u(0) = 0, \quad u(1) = 1.$$

1.9 Convert the following system to one with homogeneous boundary conditions:

$$(x^2u')' - n(n+1)u = 0, \quad u(0) = 0, \quad u(1) = 1.$$

1.10 Obtain the Green's function for the equation

$$(xu')' - \frac{9}{x}u = f(x), \quad u(0) = 0, \quad u(1) = 0.$$

Using the Green's function, find an explicit solution when $f(x) = x^n$. Check if there are any values for the integer n for which your solution does not satisfy the boundary conditions.

1.11 Find the Green's function for

$$u'' + \omega^2 u = f(x), \quad u(0) = 0, \quad u(\pi) = 0.$$

Examine the special cases $\omega = n$, $n = 0,1,\dots$.

1.12 Express the equation

$$u'' - 2u' + u = f(x), \quad u(0) = 0, \quad u(1) = 0,$$

in the self-adjoint form. Obtain the solution using the Green's function when $f(x) = e^x$.

1.13 Transform the equation

$$xu'' + 2u' = f(x); \quad u'(0) = 0, \quad u(1) = 0,$$

into the self-adjoint form. Find the Green's function and express the solution in terms of $f(x)$. State the restrictions on $f(x)$ for the solution to exist.

1.14 Find the Green's function for

$$x^2 u'' - xu' + u = f(x), \quad u(0) = 0, \quad u(1) = 0.$$

1.15 Using the self-adjoint form of the differential equation

$$x^2 u'' + 3xu' - 3u = f(x), \quad u(0) = 0, \quad u(1) = 0,$$

find the Green's function and obtain an explicit solution when $f(x) = x$.

1.16 Solve the equation

$$x^2 u'' + 3xu' = x^2, \quad u(1) = 1, \quad u(2) = 2,$$

using the Green's function.

1.17 For the problem

$$Lu = u'' + u' = 0, \quad u(0) = 0, \quad u'(1) = 0,$$

obtain the adjoint system. Solve the eigenvalue problems,

$$Lu = \lambda u, \quad L^* v = \lambda v,$$

and show that their eigenfunctions are bi-orthogonal.

1.18 By solving the nonhomogeneous problem

$$u'' = \delta_\epsilon(x - \xi), \quad u(0) = 0, \quad u(1) = 0,$$

where

$$\delta_\epsilon(x) = \begin{cases} 0, & |x| > \epsilon, \\ \frac{1}{2\epsilon}, & |x| < \epsilon, \end{cases}$$

in three parts:

(a) $0 < x < \xi - \epsilon$,

(b) $\xi - \epsilon < x < \xi + \epsilon$, and

(c) $\xi + \epsilon < x < 1$,

show that, in the limit $\epsilon \to 0$, we recover the Green's function.

1.19 Expanding

$$g(x, \xi) = \begin{cases} x(\xi - 1), & x < \xi, \\ \xi(x - 1), & x > \xi, \end{cases}$$

in terms of $u_n = \sqrt{2} \sin n\pi x$ as a Fourier series, show that

$$g(x, \xi) = \sum_{n=1}^{\infty} \frac{u_n(x) u_n(\xi)}{\lambda_n}, \quad \lambda_n = -\pi^2 n^2.$$

1.20 Obtain the Green's function for

$$\kappa \nabla^2 u(x_1, x_2) = v_1 \frac{\partial u}{\partial x_1} + v_2 \frac{\partial u}{\partial x_2},$$

where v_1 and v_2 are constants, by transforming the dependent variable.

1.21 The anisotropic Laplace equation in a two-dimensional infinite domain is given by

$$k_1^2 \frac{\partial^2 u}{\partial x_1^2} + k_2^2 \frac{\partial^2 u}{\partial x_2^2} = 0.$$

Find the Green's function for this equation.

1.22 In a semi-infinite medium, $-\infty < x < \infty, 0 < y < \infty$, the pressure fluctuations satisfy the wave equation

$$\nabla^2 p = \frac{1}{c^2} \frac{\partial^2 p}{\partial t^2},$$

where c is the wave speed and t is time. If the boundary, $y = 0$, is subjected to a pressure $p = P_0 \delta(x) h(t)$, show that

$$p(x,y,t) = A \frac{y}{r^2} \frac{t}{\sqrt{t^2 - k^2 r^2}} h(t - kr),$$

where $r = \sqrt{x^2 + y^2}$ and $k = 1/c$, is a solution of the wave equation. Evaluate the constant A using equilibrium of the medium in the neighborhood of the applied load. Hint: Use polar coordinates.

1.23 For the two-dimensional wave equation

$$\nabla^2 u = \frac{1}{c^2} \frac{\partial^2 u}{\partial t^2},$$

steady-state solutions are obtained using $u(x,y,t) = v(x,y) e^{-i\Omega t}$, where v satisfies the Helmholtz equation,

$$Lv = 0, \quad L = \nabla^2 + k^2, \quad k = \Omega/c.$$

Show that the Hankel functions $H_0^{(1)}(kr)$ and $H_0^{(2)}(kr)$ satisfy $Lg = \delta(x,y)$ when $r = \sqrt{x^2 + y^2} \neq 0$. Examine their asymptotic forms for $kr << 1$ and for $kr >> 1$, using the results shown in Abramowitz and Stegun (1965) and select multiplication constants A and B to have $g = AH_0^{(1)}$ or $g = BH_0^{(2)}$ by comparing the asymptotic form with the Green's function for the Laplace operator ($k \to 0$). Show that $H_0^{(1)}$ corresponds to an outgoing wave and $H_0^{(2)}$ to an incoming wave.

1.24 Show that

$$g = -\frac{e^{\pm ikr}}{4\pi r},$$

satisfies the Helmholtz equation

$$\nabla^2 g + k^2 g = \delta(x - \xi),$$

in a 3D infinite domain, with $r = |x - \xi|$. Assuming the Helmholtz equation is obtained from the wave equation by separating the time dependence using a factor $e^{-i\Omega t}$, show that ($\pm$) signs correspond to outgoing and incoming waves, respectively.

1.25 Consider a volume V enclosed by the surface S in 3D. From

$$\nabla^2 u + k^2 u = f, \quad \nabla^2 g + k^2 g = \delta(x - \xi),$$

obtain the solution

$$u(\xi) = \int_V g(\xi, x) f(x) dV - \int_S \left[g \frac{\partial u}{\partial n} - u \frac{\partial g}{\partial n} \right] dS.$$

1.26 In the previous problem, assuming V is a sphere of radius R centered at $x = 0$ and x is a point on its surface, and g is the Green's function for the 3D infinite space, show that

$$u(\xi) = \frac{1}{4\pi} \int_S \left[\frac{e^{ikr}}{r} \frac{\partial u}{\partial n} - u \frac{\partial}{\partial n} \left(\frac{e^{ikr}}{r} \right) \right] dS,$$

if $f = 0$. If the included angle between x and $x - \xi$ is ψ, show that $dr/dn = \cos \psi$. Also show that as $R \to \infty$, $r \to R$ and $R^2 (1 - \cos \psi)$ is finite and for the surface integral to exist

$$r \left(\frac{\partial u}{\partial r} - iku \right) \to 0, \quad \text{and} \quad u \to 0.$$

These are known as the Sommerfeld radiation conditions.

1.27 By reconsidering the previous problem, show that the surface integral exists if the less restricted condition

$$r \left[\frac{\partial u}{\partial r} - iku + \frac{u}{r} \right] \to 0$$

is satisfied.

1.28 A wedge-shaped 2D domain has the boundaries: $y = 0$ and $y = x$. Obtain the Green's function for the Poisson equation, $\nabla^2 u = f(x, y)$, for this domain if $u(x, 0) = 0$ and $u(x, x) = 0$. Use the method of images for your solution.

1.29 A unit circle is mapped into a *cardioid* by the transformation

$$w = c(1+z)^2,$$

where c is a real number. Obtain the Green's function, g, for a domain in the shape of a cardioid with $g = 0$ on the boundary.

1.30 The steady-state temperature in a semi-infinite plate satisfies

$$\nabla^2 u(x,y) = 0, \quad -\infty < x < \infty, \quad 0 < y < \infty.$$

On the boundary, $y = 0$, the temperature is given as $u = f(x)$. Obtain the temperature distribution in the domain using the Green's function.

1.31 Show by substitution that the solutions of

$$[(1-x^2)u']' - \frac{h^2}{1-x^2}u = 0$$

are

$$u = \left(\frac{1-x}{1+x}\right)^{\pm h/2}.$$

1.32 Obtain the Green's function for the above operator when $h \neq 0$ if the boundary conditions are

$$u(-1) = \text{finite}, \quad u(1) = \text{finite}.$$

1.33 If $h = 0$ in the preceding problem, show that

$$U(x) = \frac{1}{\sqrt{2}}$$

is a normalized solution of the above homogeneous equation, which satisfies both the boundary conditions. Obtain the generalized Green's function for this case.

1.34 Find the generalized Green's function for

$$u'' = f(x), \quad u(1) = u(-1), \quad u'(1) = u'(-1).$$

1.35 To solve the self-adjoint problem

$$Lu = f(x),$$

with homogeneous boundary conditions at $x = a$ and $x = b$, we use a generalized Green's function g which satisfies

$$Lg = \delta(x - \xi) - U(x)U(\xi),$$

with homogeneous boundary conditions, where U is the normalized solution of the homogeneous equation satisfying the same boundary conditions. The solution is written as

$$u(x) = AU(x) + \int_a^b g(x,\xi)f(\xi)\,d\xi.$$

By operating on this equation using L, show that u is the required solution.

1.36 For the equation

$$Lu = u'' + u' = 0, \quad u'(0) = u'(1) = 0,$$

obtain the adjoint equation and its boundary conditions. Obtain normalized homogeneous solutions of $LU = 0$ and $L^*U^* = 0$. Construct generalized Green's functions g and g^*.

2

INTEGRAL EQUATIONS

An equation involving the integral of an unknown function is called an **integral equation**. The unknown function itself may appear explicitly in an integral equation along with its integral. If the derivatives of the unknown are also present, we have what are known as integro-differential equations. Here are some examples of integral equations:

$$u(x) - \int_0^1 (1 + x\xi)u(\xi)\,d\xi = x^2, \tag{2.1}$$

$$\int_0^\pi \sin(x + \xi)u(\xi)\,d\xi = \cos(x), \tag{2.2}$$

$$\int_0^x \frac{u(\xi)\,d\xi}{\sqrt{x - \xi}} = f(x). \tag{2.3}$$

2.1 CLASSIFICATION

One way of classifying integral equations is based on the explicit presence of the unknown function. Integral equations of the **first kind** do not have the unknown function present and these equations have the form

$$\int_{a(x)}^{b(x)} k(x,\xi)u(\xi)\,d\xi = f(x), \tag{2.4}$$

where $k(x,\xi)$ is called the kernel of the integral equation. If

$$k(x,\xi) = k(\xi,x), \tag{2.5}$$

the kernel is called a **symmetric** kernel.

An integral equation of the **second kind** will have the unknown explicitly in it, for example

$$u(x) - \int_{a(x)}^{b(x)} k(x,\xi)u(\xi)\,d\xi = f(x). \tag{2.6}$$

The integral equations of the first and second kind are further classified as **Fredholm type** and **Volterra type** based on the nature of the limits of integration. If the limits are constants, such as

$$u(x) - \int_{a}^{b} k(x,\xi)u(\xi)\,d\xi = f(x), \tag{2.7}$$

we have a Fredholm equation of the second kind. If the upper or lower limit depends on the independent variable, x, we have Volterra equations.

Unlike differential equations, there are no separate conditions such as the boundary conditions in the integral equation formulations. In certain problems involving infinite domains, based on the physics, conditions on the growth of the unknown as $|x| \to \infty$ may be imposed.

We assume the kernel, $k(x,\xi)$, is continuous in x and ξ. The Volterra equation,

$$u(x) - \int_{a}^{x} k(x,\xi)u(\xi)\,d\xi = f(x), \tag{2.8}$$

may be written in the Fredholm form,

$$u(x) - \int_{a}^{b} \bar{k}(x,\xi)u(\xi)\,d\xi = f(x), \tag{2.9}$$

provided

$$\bar{k}(x,\xi) = \begin{cases} k(x,\xi), & \xi < x, \\ 0, & \xi > x. \end{cases} \tag{2.10}$$

The continuity of the kernel now requires,

$$k(x,x) = 0. \tag{2.11}$$

Integral equations with discontinuous kernels are called **singular integral equations**. We will consider these later in this chapter. In the preceding equations, if the forcing function $f(x) = 0$, the equations are homogeneous. The foregoing examples are all linear integral equations. For homogeneous, linear integral equations, if u_1 and u_2 are solutions, $u_1 + u_2$ and cu_1 are also solutions. Here c is any constant.

In higher dimensions, with

$$x = x_1, x_2, \ldots, x_n, \quad \xi = \xi_1, \xi_2, \ldots, \xi_n, \quad \mathbf{d\xi} = d\xi_1 d\xi_2 \ldots d\xi_n, \quad (2.12)$$

we can have

$$u(x) - \int_\Omega k(x, \xi) u(\xi) \, \mathbf{d\xi} = f(x), \quad (2.13)$$

where Ω is the domain of integration. Using u and f as vector-valued functions and k as a matrix function, we may also extend this to a coupled system of integral equations.

2.2 INTEGRAL EQUATION FROM DIFFERENTIAL EQUATIONS

It is always possible to convert a differential equation into an integral equation. However, in general, it is not possible to convert an integral equation into a differential equation.

Consider a differential equation of the form

$$Lu = h, \quad (2.14)$$

with homogeneous boundary conditions. If we know the Green's function for the operator, L, we may formally write a solution for u in terms of the forcing function, h. As we know, this is not always possible. However, we may split the operator, L, as

$$L = L_1 + L_2, \quad (2.15)$$

where we know the Green's function, $g_1(x,\xi)$, for L_1. Then,

$$L_1 u = h - L_2 u, \quad u = \int_a^b g_1(x,\xi)[h - L_2 u]\,d\xi. \qquad (2.16)$$

This is now an integral equation,

$$u - \int_a^b k(x,\xi)u(\xi)\,d\xi = f(x), \qquad (2.17)$$

where

$$k(x,\xi) = -g_1(x,\xi)L_2, \quad f(x) = \int_a^b g_1(x,\xi)h(\xi)\,d\xi. \qquad (2.18)$$

If L_2 consists of algebraic terms, we are done with the conversion. If it has certain differential operators, we may need some integration by parts.

2.3 EXAMPLE: CONVERTING DIFFERENTIAL EQUATION

Consider the differential equation,

$$u'' + x^n u = h(x), \quad u(0) = 0, \quad u(1) = 0. \qquad (2.19)$$

We cannot find the Green's function for the operator

$$L = \frac{d^2}{dx^2} + x^n, \qquad (2.20)$$

but we know the Green's function for the partial operator, d^2/dx^2,

$$g(x,\xi) = \begin{cases} x(\xi - 1), & \xi < x, \\ \xi(x - 1), & \xi > x. \end{cases} \qquad (2.21)$$

Using this we get the integral equation,

$$u(x) + \int_0^1 g(x,\xi)\xi^n u(\xi)\,d\xi = f(x), \qquad (2.22)$$

where

$$f(x) = \int_0^1 g(x,\xi)h(\xi)\,d\xi. \qquad (2.23)$$

2.4 SEPARABLE KERNEL

The kernel, $k(x,\xi)$, is called separable (or degenerate) if it can be written as

$$k(x,\xi) = \sum_{n=1}^{N} g_n(x)h_n(\xi),\qquad(2.24)$$

where N is finite.

As an example, consider

$$u(x) - \int_0^1 (1+x\xi)u(\xi)\,d\xi = f(x).\qquad(2.25)$$

Using

$$c_1 = \int_0^1 u(\xi)d\xi, \quad c_2 = \int_0^1 \xi u(\xi)\,d\xi,\qquad(2.26)$$

we have

$$u(x) = f(x) + c_1 + c_2 x.\qquad(2.27)$$

Using the defining relations, Eq. (2.26), from the preceding expression we have

$$c_1 = \langle f,1\rangle + c_1 + \frac{1}{2}c_2,$$
$$c_2 = \langle f,x\rangle + \frac{1}{2}c_1 + \frac{1}{3}c_2,\qquad(2.28)$$

where we have used the notation

$$\langle f,g\rangle = \int_0^1 fg\,dx.\qquad(2.29)$$

The system of simultaneous equations,

$$-\frac{1}{2}c_2 = \langle f,1\rangle,$$
$$-\frac{1}{2}c_1 + \frac{2}{3}c_2 = \langle f,x\rangle,$$

can be solved to get

$$c_2 = -2\langle f,1\rangle,$$
$$c_1 = -2\langle f,x\rangle - \frac{8}{3}\langle f,1\rangle.\qquad(2.30)$$

The solution of the integral equation is

$$u(x) = f(x) - 2\langle f, x \rangle - \frac{8}{3}\langle f, 1 \rangle - 2\langle f, 1 \rangle x. \qquad (2.31)$$

This can also be written as

$$u(x) = f(x) - \frac{1}{3}\int_0^1 [6(x+\xi) + 8]f(\xi)\, d\xi. \qquad (2.32)$$

If $f(x)$ is orthogonal to the functions 1 and x, this reduces to simply

$$u(x) = f(x). \qquad (2.33)$$

Now we are ready to look at the general case of a separable kernel with N terms. From

$$u(x) - \int_a^b \sum_{n=1}^N g_n(x) h_n(\xi) u(\xi)\, d\xi = f(x), \qquad (2.34)$$

we identify

$$c_n = \int_a^b h_n(\xi) u(\xi)\, d\xi. \qquad (2.35)$$

Then Eq. (2.34) becomes

$$u(x) - \sum_{n=1}^N c_n g_n(x) = f(x). \qquad (2.36)$$

Multiplying this with $h_m(x)$ and integrating, we find

$$c_m - \sum_{n=1}^N \left[\int_a^b h_m(x) g_n(x)\, dx \right] c_n = \int_a^b f(x) h_m(x)\, dx, \quad m = 1, 2, \ldots, N,$$

where we used the definition, Eq. (2.35). This system has a matrix representation,

$$[\mathbf{I} - \mathbf{A}]\mathbf{c} = \mathbf{b}, \qquad (2.37)$$

where $\mathbf{I}$ is the identity matrix, $\mathbf{c}$ is a column of c_n, and the elements of $\mathbf{A}$ and $\mathbf{b}$ are given by

$$A_{mn} = \int_a^b h_m(x) g_n(x)\, dx, \quad b_m = \int_a^b f(x) h_m(x)\, dx. \qquad (2.38)$$

If the matrix, $\mathbf{I} - \mathbf{A}$, is not singular, we can invert it and solve for the constants, c_n.

From the series representation of the separable kernel, Eq. (2.24), it is not clear how to identify a symmetric kernel. For a given term $g_n(x)h_n(\xi)$, if there is another term $g_m(x)h_m(\xi)$ for some unique value of m, such that

$$g_n(x)h_n(\xi) = g_m(\xi)h_m(x), \quad \text{or} \quad g_n(x) = h_m(x), \qquad (2.39)$$

we have a symmetric kernel. In this case, the matrix, $\mathbf{A}$, will be symmetric. Another way to look at this is by expanding g_m and h_m using N orthogonal functions f_n. Then k has the form

$$k(x,\xi) = \sum_{i=1}^{N}\sum_{j=1}^{N} c_{ij}f_i(x)f_j(\xi). \qquad (2.40)$$

Symmetry of the kernel requires $c_{ij} = c_{ji}$.

2.5 EIGENVALUE PROBLEM

As in the case of matrices and linear differential operators, the integral operator may transform a function into a scalar multiple of itself. This relation is written as

$$u(x) = \lambda \int_a^b k(x,\xi)u(\xi)\,d\xi, \qquad (2.41)$$

where λ is the eigenvalue and u is the eigenfunction. The special cases, $\lambda = 0$, which leads to $u(x) = 0$, and $\lambda = \infty$, which leads to the orthogonality of u and k are excluded. This statement of the eigenvalue problem is in agreement with differential eigenvalue problems if we examine

$$Lu = \lambda u, \quad u = \lambda L^{-1}u = \lambda \int g(x,\xi)u(\xi)\,d\xi. \qquad (2.42)$$

2.5.1 Example: Eigenvalues

Consider

$$u(x) = \lambda \int_0^\pi \sin(x+\xi)u(\xi)\,d\xi. \tag{2.43}$$

Expanding $\sin(x+\xi)$ as

$$\sin(x+\xi) = \sin x \cos \xi + \cos x \sin \xi, \tag{2.44}$$

we get

$$u(x) = \lambda[c_1 \sin x + c_2 \cos x], \tag{2.45}$$

where

$$c_1 = \int_0^\pi \cos \xi u(\xi)\,d\xi, \quad c_2 = \int_0^\pi \sin \xi u(\xi)\,d\xi. \tag{2.46}$$

Multiplying Eq. (2.45) by $\cos x$ and $\sin x$, respectively, and integrating, we find

$$c_1 = \frac{\lambda \pi}{2}c_2, \quad c_2 = \frac{\lambda \pi}{2}c_1. \tag{2.47}$$

The determinant of this system gives the characteristic equation,

$$1 - \frac{\lambda^2 \pi^2}{4} = 0. \tag{2.48}$$

The values of λ are

$$\lambda_1 = -2/\pi, \quad \lambda_2 = 2/\pi. \tag{2.49}$$

Substituting these in Eq. (2.47), we get

$$\lambda = \lambda_1, \quad c_1 = -c_2; \quad \lambda = \lambda_2, \quad c_1 = c_2. \tag{2.50}$$

The corresponding eigenfunctions are

$$u_1 = \sin x - \cos x, \quad u_2 = \sin x + \cos x. \tag{2.51}$$

As these are solutions of a homogeneous equation, we are free to choose $c_1 = 1$. If we need normalized solutions, we set $c_1 = 1/\sqrt{\pi}$.

2.5.2 Nonhomogeneous Equation with a Parameter

Let us consider the same kernel in a nonhomogeneous equation,

$$u(x) = \lambda \int_0^\pi \sin(x+\xi)u(\xi)\,d\xi + f(x), \qquad (2.52)$$

where λ is a given parameter.

Expanding $\sin(x+\xi)$ as before, we have

$$u(x) = \lambda[c_1 \sin x + c_2 \cos x] + f(x), \qquad (2.53)$$

where

$$c_1 = \int_0^\pi \cos\xi\, u(\xi)\,d\xi, \quad c_2 = \int_0^\pi \sin\xi\, u(\xi)\,d\xi. \qquad (2.54)$$

Multiplying Eq. (2.53) by $\cos x$ and $\sin x$, respectively, and integrating, we find

$$c_1 - \frac{\lambda\pi}{2}c_2 = \langle f,\cos\rangle, \quad -\frac{\lambda\pi}{2}c_1 + c_2 = \langle f,\sin\rangle, \qquad (2.55)$$

where the dummy variable in side the inner products have been suppressed. If the determinant of this system is not zero, we find

$$\begin{aligned}
c_1 &= \frac{1}{1-\lambda^2\pi^2/4}\int_0^\pi \left[\cos\xi + \frac{\lambda\pi}{2}\sin\xi\right]f(\xi)\,d\xi, \\
c_2 &= \frac{1}{1-\lambda^2\pi^2/4}\int_0^\pi \left[\sin\xi + \frac{\lambda\pi}{2}\cos\xi\right]f(\xi)\,d\xi.
\end{aligned} \qquad (2.56)$$

The determinant becomes zero when λ happens to be the eigenvalues, $\pm 2/\pi$, of the operator. For example, when $\lambda = -2/\pi$, Eq. (2.55) shows

$$c_1 + c_2 = \langle f,\cos\rangle, \quad c_1 + c_2 = \langle f,\sin\rangle. \qquad (2.57)$$

Subtracting the first equation from the second, we have

$$\int_0^\pi [\sin x - \cos x]f(x)\,dx = 0. \qquad (2.58)$$

If this orthogonality of f with the eigenfunction u_1 is not satisfied, the solution for $\lambda = \lambda_1$ does not exist.

If the orthogonality is satisfied, we get

$$c_2 = -c_1 + \langle f, \cos \xi \rangle, \tag{2.59}$$

where c_1 remains as an unknown. We, then, have the nonunique solution,

$$u = f(x) - \frac{2}{\pi} \left[\int_0^\pi f(\xi) \cos \xi \cos x \, d\xi + c_1 u_1 \right]. \tag{2.60}$$

The situation when $\lambda = \lambda_2$ can be dealt with along similar lines.

2.6 HILBERT-SCHMIDT THEORY

An integral equation with a symmetric kernel has properties analogous to those of symmetric matrices and self-adjoint differential equations. There are two important theorems pertaining to the eigenvalue problem,

$$u(x) = \lambda \int_a^b k(x,\xi) u(\xi) \, d\xi, \tag{2.61}$$

where k is real symmetric.

I. If two eigenvalues λ_m and λ_n are distinct, the corresponding eigenfunctions u_m and u_n are orthogonal,
II. All eigenvalues are real.

To prove these, consider, for $\lambda_m \neq \lambda_n$,

$$u_m(x) = \lambda_m \int_a^b k(x,\xi) u_m(\xi) \, d\xi, \quad u_n(x) = \lambda_n \int_a^b k(x,\xi) u_n(\xi) \, d\xi. \tag{2.62}$$

Since zeros as eigenvalues are not allowed, we rewrite these as

$$\int_a^b k(x,\xi) u_m(\xi) \, d\xi = \frac{u_m(x)}{\lambda_m}, \quad \int_a^b k(x,\xi) u_n(\xi) \, d\xi = \frac{u_n(x)}{\lambda_n}. \tag{2.63}$$

Next, we multiply the first equation by u_n and the second by u_m and integrate and subtract to get

$$\int_a^b \int_a^b [u_n(x)k(x,\xi)u_m(\xi) - u_m(x)k(x,\xi)u_n(\xi)]\,d\xi\,dx$$

$$= \left(\frac{1}{\lambda_m} - \frac{1}{\lambda_n}\right) \langle u_m, u_n \rangle. \tag{2.64}$$

By interchanging the variables, x and ξ, in the first term inside the integral,

$$\int_a^b \int_a^b u_n(\xi)[k(\xi,x) - k(x,\xi)]u_m(x)\,d\xi\,dx = \left(\frac{1}{\lambda_m} - \frac{1}{\lambda_n}\right) \langle u_m, u_n \rangle. \tag{2.65}$$

As $k(x,\xi) = k(\xi,x)$, we obtain the orthogonality relation

$$\langle u_m, u_n \rangle = 0. \tag{2.66}$$

To show that the eigenvalues are real, we begin by assuming the contrary. That is, assume the eigenvalues are complex and hope this assumption would lead to an absurd result. Let

$$\lambda_m = \alpha + i\beta. \tag{2.67}$$

Since k is real, $\alpha - i\beta$, the complex conjugate of λ_m should also be an eigenvalue. Let

$$\lambda_n = \alpha - i\beta.$$

The two eigenfunctions are now complex. Let

$$u_m = v + iw, \quad u_n = v - iw.$$

Using these on the right-hand side of Eq. (2.65), we get

$$-2i\beta \int_a^b [v^2 + w^2]\,dx = 0.$$

In this equation, the integrand is positive, and, then, the integral is non-zero. Thus, β must be zero. Thus, our original assumption about complex eigenvalues leads to a contradiction.

If all the eigenvalues of k are positive, k is called a **positive definite kernel**.

2.7 ITERATIONS, NEUMANN SERIES, AND RESOLVENT KERNEL

For a given Fredholm equation of the second kind,

$$u(x) = f(x) + \lambda \int_a^b k(x,\xi)u(\xi)\,d\xi, \tag{2.68}$$

we may assume, as a first approximation,

$$u^{(0)} = f(x). \tag{2.69}$$

Using this in Eq. (2.68), we find as the result of first iteration,

$$u^{(1)} = f(x) + \lambda \int_a^b k(x,\xi)f(\xi)\,d\xi. \tag{2.70}$$

Introducing this into Eq. (2.68), we get

$$u^{(2)} = f(x) + \lambda \int_a^b k(x,\xi)f(\xi)\,d\xi + \lambda^2 \int_a^b \int_a^b k(x,\eta)k(\eta,\xi)f(\xi)\,d\eta\,d\xi. \tag{2.71}$$

It is convenient to define **iterated kernels**,

$$k^{(2)}(x,\xi) = \int_a^b k(x,\eta)k(\eta,\xi)\,d\eta, \tag{2.72}$$

$$k^{(n)} = \int_a^b k^{(n-1)}(x,\eta)k(\eta,\xi)\,d\eta. \tag{2.73}$$

We may add k itself into this group, in the form

$$k^{(1)}(x,\xi) = k(x,\xi). \tag{2.74}$$

With the preceding notation, the result of the Nth iteration can be found as

$$u^{(N)}(x) = f(x) + \lambda \int_a^b \sum_{n=1}^{N} \lambda^{(n-1)} k^{(n)}(x,\xi)f(\xi)\,d\xi. \tag{2.75}$$

As $N \to \infty$, assuming the sum converges, we have

$$u(x) = f(x) + \lambda \int_a^b g(x,\xi) f(\xi) \, d\xi, \qquad (2.76)$$

where

$$g(x,\xi) = \lim_{N \to \infty} \sum_{n=1}^N \lambda^{(n-1)} k^{(n)}(x,\xi). \qquad (2.77)$$

Here, the series on the right–hand side is called a **Neumann series**, and $g(x,\xi)$ is called the **resolvent kernel**.

2.7.1 Example: Neumann Series

Consider the equation

$$u(x) = f(x) + \lambda \int_0^\pi \sin(x+\xi) u(\xi) \, d\xi. \qquad (2.78)$$

Here,

$$k^{(1)} = k(x,\xi) = \sin x \cos \xi + \cos x \sin \xi. \qquad (2.79)$$

The iterated kernels may be computed using

$$\int_0^\pi \cos \eta \sin \eta \, d\eta = 0, \quad \int_0^\pi \cos^2 \eta \, d\eta = \frac{\pi}{2}, \quad \int_0^\pi \sin^2 \eta \, d\eta = \frac{\pi}{2}, \quad (2.80)$$

as

$$k^{(2)} = \int_0^\pi [\sin x \cos \eta + \cos x \sin \eta][\sin \eta \cos \xi + \cos \eta \sin \xi] \, d\eta$$

$$= \frac{\pi}{2}[\sin x \sin \xi + \cos x \cos \xi] = \frac{\pi}{2} \cos(x - \xi), \qquad (2.81)$$

$$k^{(3)} = \int_0^\pi \frac{\pi}{2}[\sin x \sin \eta + \cos x \cos \eta][\sin \eta \cos \xi + \cos \eta \sin \xi] \, d\eta$$

$$= \frac{\pi^2}{4}[\sin x \cos \xi + \cos x \sin \xi] = \frac{\pi^2}{4} \sin(x + \xi). \qquad (2.82)$$

We notice that $k^{(1)}$, $k^{(3)}$, $k^{(5)}$, etc., contain $\sin(x+\xi)$ and $k^{(2)}$, $k^{(4)}$, $k^{(6)}$, etc., contain $\cos(x-\xi)$. The Neumann series can be written as

$$g(x,\xi) = \left(1 + \frac{\lambda^2\pi^2}{4} + \frac{\lambda^4\pi^4}{16} + \cdots\right)\left[\sin(x+\xi) + \frac{\lambda\pi}{2}\cos(x-\xi)\right].$$

(2.83)

When $|\lambda| < 2/\pi$, we may sum the series to find

$$g(x,\xi) = \frac{1}{1 - \lambda^2\pi^2/4}[\sin(x+\xi) + \frac{\lambda\pi}{2}\cos(x-\xi)].$$

(2.84)

The following example shows, in fact, this expression is valid even when $|\lambda| > 2/\pi$.

2.7.2 Example: Direct Calculation of the Resolvent Kernel

From the equation

$$u(x) = \lambda\int_0^\pi [\sin x\cos\xi + \cos x\sin\xi]u(\xi)\,d\xi + f(x),$$

(2.85)

by defining

$$c_1 = \langle f,\cos\rangle, \quad c_2 = \langle f,\sin\rangle,$$

(2.86)

we have

$$u(x) = \lambda[c_1\sin x + c_2\cos x] + f(x).$$

(2.87)

Multiplying this by $\cos x$ and integrating, we find

$$c_1 = \frac{\lambda\pi}{2}c_2 + \langle f,\cos\rangle.$$

(2.88)

Similarly, using $\sin x$,

$$c_2 = \frac{\lambda\pi}{2}c_1 + \langle f,\sin\rangle.$$

(2.89)

Solving for c_1 and c_2,

$$c_1 = \frac{1}{1 - \lambda^2\pi^2/4}[\langle f,\cos\rangle + \frac{\lambda\pi}{2}\langle f,\sin\rangle],$$

(2.90)

$$c_2 = \frac{1}{1-\lambda^2\pi^2/4}[\langle f,\sin\rangle + \frac{\lambda\pi}{2}\langle f,\cos\rangle]. \qquad (2.91)$$

Substituting these in the expression (2.87),

$$u(x) = f(x) + \frac{\lambda}{1-\lambda^2\pi^2/4}\int_0^\pi [\sin(x+\xi) + \frac{\lambda\pi}{2}\cos(x-\xi)]f(\xi)\,d\xi. \qquad (2.92)$$

The resolvent kernel can be seen as

$$g(x,\xi) = \frac{1}{1-\lambda^2\pi^2/4}[\sin(x+\xi) + \frac{\lambda\pi}{2}\cos(x-\xi)], \qquad (2.93)$$

which is valid for all values of λ, except when λ is an eigenvalue, (i.e., $\lambda = \pm 2/\pi$).

2.8 QUADRATIC FORMS

Associated with a symmetric kernel $k(x,\xi)$, we have the quadratic form

$$J[u] = \int_a^b \int_a^b k(x,\xi)u(x)u(\xi)d\xi\,dx. \qquad (2.94)$$

If the functions u are selected from the set satisfying $\|u\| = 1$, the extremum values of J can be found by setting the first variation of the modified functional

$$J^*[u] = J[u] - \mu\left[\int_a^b u^2 dx - 1\right] \qquad (2.95)$$

to zero. Here, μ is a Lagrange multiplier. Using the calculus of variations, we obtain the eigenvalue problem

$$u(x) = \lambda \int_a^b k(x,\xi)u(\xi)d\xi, \qquad (2.96)$$

where $\lambda = 1/\mu$. For a particular normalized eigenfunction u_m, Eq. (2.96) becomes

$$u_m(x) = \lambda_m \int_a^b k(x,\xi)u_m(\xi)d\xi. \qquad (2.97)$$

Multiplying both sides of Eq. (2.97) by $u_m(x)$ and integrating, we find

$$1 = \lambda_m J[u_m]. \tag{2.98}$$

This shows that the extremum values of J corresponding to the eigenfunctions satisfy

$$J[u_m] = \frac{1}{\lambda_m}. \tag{2.99}$$

2.9 EXPANSION THEOREMS FOR SYMMETRIC KERNELS

We state two expansion theorems without proof:

I. If $k(x,\xi)$ is degenerate or if it has a finite number of eigenvalues of one sign (e.g., an infinite number of positive eigenvalues and a finite number of negative ones), it has the expansion

$$k(x,\xi) = \sum_i^N \frac{u_i(x)u_i(\xi)}{\lambda_i}, \tag{2.100}$$

where $N \to \infty$ when there are infinite eigenvalues.
This is known as Mercer's theorem.

II. For general symmetric kernels that are uniformly continuous in their variables x and ξ, the relation

$$f(x) = \int_a^b k(x,\xi)g(\xi)\,d\xi, \tag{2.101}$$

where g is a piecewise continuous function, can be considered as an integral transformation of g with respect to the kernel k. Functions f obtained in this way can be expanded in terms of the eigenfunctions of k in the form

$$f(x) = \sum_i^\infty a_i u_i(x). \tag{2.102}$$

For the proofs of Eqs. (2.100) and (2.102), the reader may consult Courant and Hilbert (1953). The second expansion theorem shows that the iterated kernel $k^{(2)}(x,\xi)$ has the form

$$k^{(2)}(x,\xi) = \sum_{i}^{\infty} a_i u_i(x). \tag{2.103}$$

2.10 EIGENFUNCTIONS BY ITERATION

Solutions of the eigenvalue problem

$$u(x) = \lambda \int_a^b k(x,\xi)u(\xi)d\xi \tag{2.104}$$

may be approximated by iterative means as follows: We choose a normalized function $u^{(0)}$ and use it on the right-hand side of Eq. (2.104). We, again, normalize the result of the integration and call it $u^{(1)}$ to use in subsequent iterations. This procedure creates a sequence of functions that would converge to an eigenfunction u_1 corresponding to the eigenvalue with smallest value of $|\lambda|$.

To see the validity of this procedure, we observe the following: By the second expansion theorem, $u^{(1)}$ will have the form

$$u^{(1)} = \sum_{i}^{\infty} a_i u_i(x). \tag{2.105}$$

The next iteration gives

$$u^{(2)} = \sum_{i}^{\infty} \frac{a_i u_i(x)}{\lambda_i} \tag{2.106}$$

and

$$u^{(n)} = \sum_{i}^{\infty} \frac{a_i u_i(x)}{\lambda_i^{(n-1)}}. \tag{2.107}$$

If λ_1 is the smallest eigenvalue magnitude, we can write

$$u^{(n)} = \frac{1}{\lambda_1^{n-1}} \left[a_1 u_1 + a_2 \left(\frac{\lambda_1}{\lambda_2} \right)^{n-1} u_2 + \cdots \right]. \tag{2.108}$$

As $n \to \infty$, all the terms except the u_1 term go to zero. The normalizations will get rid off the constants multiplying u_1. Once u_1 is known, λ_1 can be obtained from

$$\int_a^b k(x,\xi)u(\xi)\,d\xi = \frac{u_1(x)}{\lambda_1}. \tag{2.109}$$

This iterative method is known as Kellogg's method.

To use this approach for higher eigenfunctions, we must make sure our starting function is orthogonal to u_1, and after each iteration the orthogonality has to be reimposed. If the kernel satisfies the conditions for the expansion theorem of Eq. (2.100), we may work with the alternate kernel,

$$\hat{k} = k(x,\xi) - \frac{u_1(x)u_1(\xi)}{\lambda_1}, \tag{2.110}$$

to find the second eigenfunction.

2.11 BOUND RELATIONS

Let us assume the eigenvalue λ_i has M eigenfunctions u_{im} ($m = 1,2,\ldots,M$). Using the Bessel inequality with the norm of $k(x,\xi)$ with ξ and its projections on the M orthonormal eigenfunctions, we have

$$\int_a^b k^2(x,\xi)\,d\xi \geq \sum_{m=1}^{M}\left[\int_a^b k(x,\xi)u_{im}(\xi)\,d\xi\right]^2 = \sum_{m=1}^{M}\frac{u_{im}^2(x)}{\lambda_i^2}. \tag{2.111}$$

Integrating with x, we get

$$M \leq \lambda_i^2 \int_a^b \int_a^b k^2(x,\xi)\,d\xi\,dx. \tag{2.112}$$

This puts an upper bound on the number of degenerate eigenfunctions.

If we use all eigenfunctions (discarding the double index notation for degenerate ones), we have

$$\int_a^b k^2(x,\xi)\,d\xi \geq \sum_{i=1}^{N}\left[\int_a^b k(x,\xi)u_i(\xi)\,d\xi\right]^2 = \sum_{i=1}^{N}\frac{u_i^2(x)}{\lambda_i^2}. \tag{2.113}$$

Again, integrating with x, we find

$$\sum_{i=1}^{N} \frac{1}{\lambda_i^2} \le \int_a^b \int_a^b k^2(x,\xi)\,d\xi\,dx. \tag{2.114}$$

2.12 APPROXIMATE SOLUTION

There are a number of ways of finding approximate solutions of integral equations. These range from semianalytic to purely numerical approaches.

2.12.1 Approximate Kernel

We may replace the given kernel by an approximate kernel using N orthonormal functions v_i as

$$k(x,\xi) = \sum_i a_i(x)v_i(\xi), \tag{2.115}$$

where the coefficient functions are found from

$$a_i(x) = \langle k(x,\xi), v_i(\xi) \rangle. \tag{2.116}$$

Now we have an approximate degenerate kernel.

2.12.2 Approximate Solution

To solve

$$u(x) = \int_a^b k(x,\xi)u(\xi)\,d\xi + f(x), \tag{2.117}$$

we assume a functional expansion for the unknown u in the form

$$u(x) = \sum_{i=1}^{N} a_i v_i(x), \tag{2.118}$$

where $v_i(x)$ are from a chosen orthonormal sequence. The integral equation reduces to

$$\sum_{i=1}^{N} a_i \phi_i(x) = f(x), \quad \phi_i(x) \equiv v_i(x) - \int_a^b k(x,\xi) v_i(\xi) \, d\xi. \qquad (2.119)$$

A general way to solve for the unknown constants a_i is given by the Petrov-Galerkin method. If $f(x)$ happens to be a linear combination of the known functions ϕ_i, we have a unique solution for the unknowns a_i. Otherwise, the error

$$e(x) \equiv \sum_{1}^{N} a_i \phi_i - f \qquad (2.120)$$

will not be zero almost everywhere. In the Petrov-Galerkin method, we choose a sequence of N independent functions ψ_i and let the projections of the error on these functions vanish. That is,

$$\sum_{i}^{N} a_i \langle \phi_i, \psi_j \rangle = \langle f, \psi_j \rangle, \quad j = 1, 2, \ldots, N. \qquad (2.121)$$

For a symmetric kernel, we may choose $\psi_i = v_i$ to obtain a symmetric matrix problem.

2.12.3 Numerical Solution

We begin by discretizing the interval $[a,b]$ into N points, ξ_i, $i = 1, 2, \ldots, N$. Let $u_i = u(\xi_i)$. The integral with k can be approximated by the trapezoidal or the Simpson's rule for numerical integration, for example

$$\int_a^b k(x,\xi) u(\xi) \, d\xi = \sum_{1}^{N} k(x,\xi_i) w_i u_i, \qquad (2.122)$$

where w_i are the weight coefficients of the particular numerical integration scheme. For example, the trapezoidal rule has

$$w_1 = \frac{1}{2}, \quad w_2 = 1, \quad , \ldots, \qquad (2.123)$$

and Simpson's rule has

$$w_1 = \frac{2}{3}, \quad w_2 = \frac{4}{3}, \quad \dots \tag{2.124}$$

Then

$$u(x) = \sum_{1}^{N} k(x, \xi_i) w_i u_i + f(x). \tag{2.125}$$

We may use the method of collocation in which the error is made exactly zero at N points. In our case, we choose these points as x_i. With

$$A_{ij} = k(x_i, \xi_j) w_j, \quad b_i = f(x_i), \tag{2.126}$$

we get

$$[\mathbf{I} - \mathbf{A}]\mathbf{u} = \mathbf{b}, \tag{2.127}$$

where $\mathbf{u}$ and $\mathbf{b}$ are N vectors.

In all the three preceding approximation methods, integrals may be evaluated numerically if necessary.

2.13 VOLTERRA EQUATION

Introducing a parameter λ, the Volterra equation given in Eq. (2.8) can be written as

$$u(x) = \lambda \int_a^x k(x, \xi) u(\xi) \, d\xi + f(x). \tag{2.128}$$

For continuity, the kernel has to satisfy $k(x, x) = 0$.

Using iterated kernels

$$k^{(n)}(x, \xi) = \int_a^x k^{(n-1)}(x, \eta) k(\eta, \xi) \, d\eta, \quad k^{(1)}(x, \xi) = k(x, \xi), \tag{2.129}$$

the Neumann series for this Volterra equation is obtained as

$$u(x) = f(x) + \sum_{n=1}^{\infty} \lambda^n \int_a^x k^{(n)}(x, \xi) f(\xi) \, d\xi. \tag{2.130}$$

If we stipulate that k and f are bounded functions, with

$$|k| < M, \quad |f| < F, \tag{2.131}$$

we get

$$\left| \int_a^x k(x,\xi) f(\xi)\, d\xi \right| < MF(x-a)$$

and

$$\left| \int_a^x k^{(n)}(x,\xi) f(\xi)\, d\xi \right| < M^n F \frac{(x-a)^n}{n!}. \tag{2.132}$$

As $n \to \infty$, the Neumann series for u uniformly converges, and we have a unique solution for the Volterra equation.

In special cases, by differentiating both sides, a Volterra equation may be converted to a differential equation.

2.13.1 Example: Volterra Equation

The equation

$$u(x) = \int_0^x \sin(x-\xi) u(\xi)\, d\xi + \cos(x) \tag{2.133}$$

has a slowly converging iterative solution. The first five iterative solutions are shown below.

$$u^{(0)} = \cos x,$$

$$u^{(1)} = \cos x + \frac{x}{2}\sin x,$$

$$u^{(2)} = \cos x - \frac{x}{8}\{x\cos x - 5\sin x\}$$

$$u^{(3)} = \cos x - \frac{x}{48}\left\{ x\cos x + \left(-33 + x^2\right)\sin x \right\},$$

$$u^{(4)} = \cos x + \frac{x}{384}\left\{ x\left(-87 + x^2\right)\cos x + \left(279 - 14x^2\right)\sin x \right\},$$

$$u^{(5)} = \cos x + \frac{x}{3840}\left\{ 5x\left(-195 + 4x^2\right)\cos x + \left(2895 - 185x^2 + x^4\right)\sin x \right\}.$$

Noting that the kernel has the cyclic property that it returns to its initial form after two differentiations, by differentiating the integral

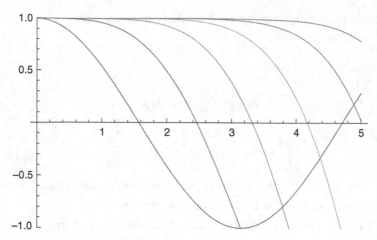

Figure 2.1. Plots of the functions $u^{(0)}$ to $u^{(5)}$ with higher iterates getting closer to the exact solution $u = 1$.

equation, we find

$$u' = \int_0^x \cos(x - \xi)u(\xi)\,d\xi - \sin x, \qquad (2.134)$$

$$u'' = 0, \qquad (2.135)$$

where we have used Eq. (2.133) to eliminate the integral of the unknown function u. We need two boundary conditions to go with this second-order equation. From Eq. (2.133), we get $u(0) = 1$. From the derivative of u, we find $u'(0) = 0$. The unique solution of the differential equation is

$$u(x) = 1. \qquad (2.136)$$

2.14 EQUATIONS OF THE FIRST KIND

We may investigate equations of the first kind,

$$\int_a^b k(x,\xi)u(\xi)\,d\xi = f(x), \qquad (2.137)$$

using the second expansion theorem stated in Section 2.9. This theorem states that if k is continuous and symmetric, f has an expansion in terms of the eigenfunctions of k. Thus, if the given f is a linear combination of the eigenfunctions, we may seek solutions. Otherwise, there is no solution. Writing

$$u = \sum_{i=1}^{N} a_i u_i(x) + \phi(x), \quad f = \sum_{i=1}^{N} b_i u_i(x), \qquad (2.138)$$

where the unknown function ϕ is orthogonal to all u_i. Using this in the integral equation, we get

$$a_i = \lambda_i b_i, \quad \int_a^b k(x,\xi)\phi(\xi)\,d\xi = 0. \qquad (2.139)$$

When there are a finite number of eigenfunctions, we find ϕ is in general nonzero; in fact, there are an infinite number of them. In conclusion, there is no unique solution for the equation of the first kind, except for the case a complete set of infinite eigenfunctions.

The Volterra equations of the first kind,

$$\int_a^x k(x,\xi)u(\xi)\,d\xi = f(x), \qquad (2.140)$$

can be converted to a Volterra equation of the second kind by differentiating both sides. We assume all the functions involved are continuous and differentiable with $k(x,x) = 0$. The first differentiation results in

$$\int_a^x \frac{\partial k}{\partial x}(x,\xi)u(\xi)\,d\xi = f'(x), \qquad (2.141)$$

and the second differentiation results in

$$\frac{\partial k}{\partial x}(x,x)u(x) + \int_a^x \frac{\partial^2 k}{\partial x^2}(x,\xi)u(\xi)\,d\xi = f''(x). \qquad (2.142)$$

Now, we have an equation of the second kind.

There are integral transform methods available for special forms of kernels. We will discuss these for the Fourier and Laplace transforms in the coming chapters.

2.15 DUAL INTEGRAL EQUATIONS

There is considerable literature on dual integral equations. The book on mixed boundary-value problems by Sneddon (1966) is a valuable resource for a deeper understanding of this topic. A knowledge of Hankel transforms is required to understand the solution techniques. Here, we will try to give a glimpse of this topic without using the Hankel transform.

Consider the axisymmetric Laplace equation in cylindrical coordinates r and z in the form

$$\frac{\partial^2 u}{\partial r^2} + \frac{1}{r}\frac{\partial u}{\partial r} + \frac{\partial^2 u}{\partial z^2} = 0, \tag{2.143}$$

where the domain is the 3D half space: $0 < z < \infty$, $0 < r < \infty$. The boundary conditions are provided in the mixed form

$$u(r,0) = u_0(r), \quad 0 < r < 1, \tag{2.144}$$

$$\frac{\partial u(r,0)}{\partial z} = 0, \quad 1 < r < \infty. \tag{2.145}$$

Mixed boundary conditions appear in many problems in physics and engineering. In electromagnetics, a charged circular plate touching the half space and in elasticity an axisymmetric body making an indentation on a surface or a circular crack in a full 3D elastic space are a few examples of these problems.

To develop solutions of the axisymmetric Laplace equation, we note the Bessel function $J_n(\rho r)$ satisfies

$$r^2 J_n'' + r J_n' + (\rho^2 r^2 - n^2) J_n = 0. \tag{2.146}$$

If we attempt a separable solution of the Laplace equation,

$$u(r,z) = v(r)e^{-\rho z}, \tag{2.147}$$

where ρ is a parameter, we find

$$r^2 v'' + rv' + \rho^2 r^2 v = 0, \tag{2.148}$$

which has

$$v = J_0(\rho r) \tag{2.149}$$

as a solution. The other solution, Y_0, is singular at $r = 0$, and we omit it. By superposition, the general solution is

$$v(r,z) = \int_0^\infty A(\rho) e^{-\rho z} J_0(\rho r) d\rho. \tag{2.150}$$

Substituting this in the mixed boundary conditions, we find

$$\int_0^\infty A(\rho) J_0(\rho r) d\rho = u_0(r), \quad 0 < r < 1, \tag{2.151}$$

$$\int_0^\infty \rho A(\rho) J_0(\rho r) d\rho = 0, \quad 1 < r < \infty. \tag{2.152}$$

So, for a certain range of r, we have one integral equation for $A(\rho)$, and for the complementary range, another integral equation. This captures the gist of dual integral equations. When $u_0 = U$, a constant, we use the following properties of the Bessel function:

$$\int_0^\infty \frac{\sin \rho}{\rho} J_0(\rho r) d\rho = \frac{\pi}{2}, \quad 0 < r < 1, \tag{2.153}$$

$$\int_0^\infty \sin \rho J_0(\rho r) d\rho = 0, \quad 1 < r < \infty, \tag{2.154}$$

to obtain

$$A(\rho) = \frac{2U}{\pi \rho} \sin \rho. \tag{2.155}$$

These properties are listed among others in Abramowitz and Stegun (1965).

2.16 SINGULAR INTEGRAL EQUATIONS

An integral equation is called singular if one of the following condition is met:

(a) The domain is infinite (semi-infinite included),
(b) The kernel k is discontinuous, and
(c) The forcing function f is discontinuous.

2.16.1 Examples: Singular Equations

$$u(x) + \int_{-\infty}^{\infty} e^{ix\xi} u(\xi)\, d\xi = f(x), \qquad (2.156)$$

$$u(x) + \int_{0}^{x} \frac{u(\xi)}{\sqrt{x - \xi}}\, d\xi = f(x), \qquad (2.157)$$

$$u(x) + \int_{-a}^{a} \sin(x\xi) u(\xi)\, d\xi = \operatorname{sgn}(x). \qquad (2.158)$$

In these examples of singular integral equations, the first one has an unbounded domain, the second has a discontinuous kernel, and the third has a discontinuous forcing function.

2.17 ABEL INTEGRAL EQUATION

The Abel integral equation is a Volterra equation of the first kind with a singular kernel, given by

$$\int_{0}^{x} \frac{u(\xi)}{\sqrt{x - \xi}}\, d\xi = f(x). \qquad (2.159)$$

Historically, this problem arose from the dynamics of a frictionless particle, under the force of gravity, sliding on a curved surface from a height x to height zero. If the time to descent, T, is given as a function, $f(x)$, of the height x, we would like to find the shape of the curve. We know the two extreme cases. For a vertical drop, the time $T = \sqrt{2x/g}$, and for a horizontal line, T is infinite. In Fig. 2.2 an arbitrary height on the curve, ξ, is taken as a parametric function of the curve length s. At the top of the curve, the particle has a potential energy of mgx and zero speed and at the height ξ, it has potential energy $mg\xi$ and kinetic energy $mv^2/2$. From the conservation of energy,

$$v^2 = 2g(x - \xi), \quad v = \sqrt{2g(x - \xi)}. \qquad (2.160)$$

As s decreases with time,

$$\frac{ds}{dt} = -v = -\sqrt{2g(x - \xi)}. \qquad (2.161)$$

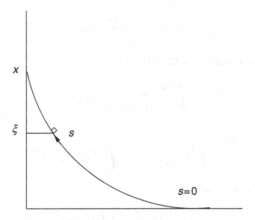

Figure 2.2. A particle sliding along a curve.

From this,

$$T(x) = \int_0^T dt = \int_{\xi=0}^x \frac{ds}{d\xi} \frac{d\xi}{\sqrt{2g(x-\xi)}}. \qquad (2.162)$$

If we set

$$u(\xi) = \frac{ds}{d\xi}, \quad f(x) = \sqrt{2g}\,T(x), \qquad (2.163)$$

we obtain the Abel equation.

This equation may be solved by multiplying both sides by $1/\sqrt{\eta-x}$ and integrating.

$$\int_0^\eta \int_0^x \frac{u(\xi)d\xi}{\sqrt{(\eta-x)(x-\xi)}}\,dx = \int_0^\eta \frac{f(x)dx}{\sqrt{\eta-x}}. \qquad (2.164)$$

Interchanging the order of integration, we have

$$\int_0^\eta u(\xi) \int_x^\eta \frac{dx}{\sqrt{(\eta-x)(x-\xi)}}\,d\xi = \int_0^\eta \frac{f(x)dx}{\sqrt{\eta-x}}. \qquad (2.165)$$

Let

$$I = \int_x^\eta \frac{dx}{\sqrt{(\eta-x)(x-\xi)}}. \qquad (2.166)$$

Using a new variable, t, in the form

$$t = \frac{x-\xi}{\eta-\xi}, \quad x = \xi + (\eta-\xi)t, \qquad (2.167)$$

$$I = \int_0^1 \frac{dt}{\sqrt{t(1-t)}}.$$ (2.168)

A second substitution, $t = \sin^2 \theta$, gives

$$I = \int_0^{\pi/2} \frac{2\sin\theta\cos\theta}{\sin\theta\cos\theta} d\theta = \pi.$$ (2.169)

Thus,

$$\int_0^\eta u(\xi)\, d\xi = \frac{1}{\pi} \int_0^\eta \frac{f(x)dx}{\sqrt{\eta - x}},$$ (2.170)

which can also be expressed as

$$u(x) = \frac{1}{\pi} \frac{d}{dx} \int_0^x \frac{f(\xi)d\xi}{\sqrt{x - \xi}}.$$ (2.171)

2.18 BOUNDARY ELEMENT METHOD

The Boundary Element Method (BEM) for the numerical solution of partial differential equations is also known as the boundary integral method. We begin with the premise that the unknown function u satisfies a partial differential equation inside a finite domain Ω with prescribed boundary conditions on the boundary $\partial\Omega$. If we know the Green's function for this problem, of course, we can write down the solution, and no numerical method is needed. In most practical cases, we have the Green's function for the infinite domain, Ω_∞, for either the full differential operator or for part of the operator. To avoid complications, let us concentrate on the first scenario. Let L denote the full operator whose Green's function g for the infinite domain is known. We also have the adjoint operator L^* and its Green's function g^*. If the source point is ξ and the observation point is x, from the given problem,

$$Lu(x) = f(x), \quad \text{in} \quad \Omega,$$ (2.172)

and the equation for the Green's function

$$L^*g^*(x,\xi) = \delta(x - \xi), \quad \text{in} \quad \Omega_\infty,$$ (2.173)

we form

$$\langle g^*, Lu \rangle - \langle u, L^*g^* \rangle = P[g^*, u]_{\partial\Omega}, \qquad (2.174)$$

where $P[g^*, u]_{\partial\Omega}$ is an integral over the boundary involving values of g^* and u and their derivatives normal to the boundary. The integration for the inner product is carried out over the domain Ω. Replacing Lu by f and L^*g^* by δ, we obtain

$$\langle g^*, f \rangle - \langle u, \delta(x - \xi) \rangle = P[g^*, u]_{\partial\Omega}. \qquad (2.175)$$

$$u(\xi) = \int_\Omega g^*(x, \xi) f(x) dx - P[g^*, u]_{\partial\Omega}, \qquad (2.176)$$

where it is assumed that the point ξ is inside the domain. As we will see later, if ξ is on the boundary curve, the integral

$$\int u(x) \delta(x - \xi) dx = \frac{\alpha}{2\pi} u(\xi), \qquad (2.177)$$

where α depends on the smoothness of the boundary. If at ξ the boundary has a continuously turning tangent, $\alpha = \pi$, and if it has a cusp, α is the included angle.

Using the inner product relation for the adjoint operators, Eq. (2.174), on

$$Lg(x, \xi_1) = \delta(x - \xi_1), \quad L^*g^*(x, \xi_2) = \delta(x - \xi_2), \qquad (2.178)$$

for the infinite domain, we find the symmetry relation

$$g(x, \xi) = g^*(\xi, x). \qquad (2.179)$$

We use this relation to write g^* in terms of g, to have Eq. (2.176) in the form

$$\frac{\alpha}{2\pi} u(\xi) = \int_\Omega g(\xi, x) f(x) dx - P[g, u]_{\partial\Omega}. \qquad (2.180)$$

In general, P contains the unknown u and its derivatives inside an integral over the boundary. Thus, Eq. (2.180) is an integral equation.

2.18.1 Example: Laplace Operator

The general statements of the previous discussion can be made concrete by focusing on the special case of the Laplace operator in a 2D space. Assume that we are interested in solving

$$\nabla^2 u = f(x), \quad \text{in} \quad \Omega, \tag{2.181}$$

with

$$u = \bar{u}(s), \quad \text{on} \quad \partial\Omega, \tag{2.182}$$

where s is a parameter describing the boundary curve. It is convenient to use the curve length for s. The coordinates of the point will be denoted by s. The Green's function for the Laplace operator in the infinite 2D domain is given by

$$g(\xi, x) = \frac{1}{2\pi} \log r, \tag{2.183}$$

where

$$r = |x - \xi| = \sqrt{(x - \xi)^2 + (y - \eta)^2}. \tag{2.184}$$

We have here two overlapping coordinate systems: $x = (x, y)$ and $\xi = (\xi, \eta)$. The boundary curve $\partial\Omega$ has distance elements ds in the (x, y) system and $d\sigma$ in the (ξ, η) system. Similarly, the unit normal to the boundary is given by n and v in the two systems. Using Eq. (2.180), we have the solution

$$\frac{\alpha}{2\pi} u(\xi) = \int_\Omega g(\xi, x) f(x) dx dy + \oint_{\partial\Omega} \left[u \frac{\partial g}{\partial n} - g \frac{\partial u}{\partial n} \right] ds. \tag{2.185}$$

As $u = \bar{u}$ on $\partial\Omega$, we rewrite the solution in the form

$$\frac{\alpha}{2\pi} u(\xi) = \int_\Omega g(x; \xi) f(x) dx dy + \oint_{\partial\Omega} \bar{u} \frac{\partial g}{\partial n} ds - \oint_{\partial\Omega} g \frac{\partial u}{\partial n} ds. \tag{2.186}$$

The unknown function $\partial u / \partial n$ inside the boundary integral has to be found first before we can evaluate u at an arbitrary interior point. We utilize the known function $(\bar{u})$ on the boundary, to set up an integral equation for

$$v(s) \equiv \frac{\partial u}{\partial n}. \tag{2.187}$$

Selecting a point (ξ, η) on the boundary, we may denote it by σ. The ensuing singular, Fredholm integral equation of the first kind is

$$\oint_{\partial\Omega} gv\,ds = \int_{\Omega} gf\,dx\,dy + \oint_{\partial\Omega} \bar{u}\frac{\partial g}{\partial n}\,ds - \frac{\alpha}{2\pi}\bar{u}(\sigma). \qquad (2.188)$$

Removing the factor 2π, we get

$$\oint_{\partial\Omega} v(s)\log|\sigma - s|\,ds$$

$$= \int_{\Omega} f(x)\log|\sigma - x|\,dx\,dy + \oint_{\partial\Omega} \bar{u}(s)\boldsymbol{n}.\nabla\log|\sigma - s|\,ds - \alpha\bar{u}(\sigma).$$

$$(2.189)$$

To solve this, we discretize the boundary curve as shown in Fig. 2.3 with nodes at $s = s_i, i = 1, 2, \ldots, N$, and let

$$v_i = v(\sigma_i) = v(s_i). \qquad (2.190)$$

We may also select collocation points identical to the nodes to obtain N equations for v_i by setting pointwise error to zero.

In evaluating the integrals involving $\log r$ and its derivative $1/r$, we face singular integrals. Methods for interpolating the unknown function between the nodes, the weight coefficients in the numerical integration, and the evaluation of the singular integrals in the sense of

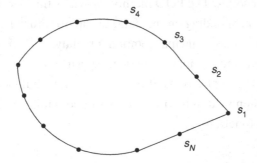

Figure 2.3. Discretized boundary in the Boundary Element Method.

Cauchy principal values are discussed in many books on this subject (see Pozrikidis, 2002, Hartmann, 1989, Brebbia, 1978).

In the example, if $\partial u/\partial n$ is given on the boundary, u on the boundary will be unknown. To solve for u, we move the interior point to the boundary and obtain a Fredholm equation of the second kind.

Compared to other numerical methods such as the Finite Element Method (FEM) and the Finite Difference Method (FDM), the number of unknowns in the system of equations in the Boundary Element Method is equal to the number of discrete points on the boundary and not in the domain. In the 2D case, the Boundary Element Method has N equations and the other two methods may have up to N^2 equations. Thus, highly accurate solutions can be obtained by BEM using minimal computer time. On the negative side, BEM requires considerable analysis from the user, and it is not readily available in the form of an all-purpose software.

2.19 PROPER ORTHOGONAL DECOMPOSITION (POD)

Proper orthogonal decomposition (POD) is a method for approximating data distributed in time and space using a finite number of orthogonal basis functions of the space variables, which are selected to minimize the expectation value of the standard deviation of the error. Generally, we are used to selecting basis functions *a prior* such as in a Fourier series. The POD method provides the best sequence of basis functions depending on the data. In the literature POD appears under various names: principal component analysis, Karhunen-Loève transformation, and singular value decomposition.

Although the applications of this method in turbulence and other dynamical systems often involve vector valued data, we discuss POD using a scalar function

$$v = v(x,t), \quad x \quad \text{in} \quad \Omega, \quad t \quad \text{in} \quad [0,T]. \tag{2.191}$$

We define the spatial norm, spatial inner product, and the temporal expectation value by

$$\|v\|^2 = \int_\Omega v^2 dx, \quad \langle v,u \rangle = \int_\Omega vu dx, \quad \mathbb{E}\{v\} = \frac{1}{T}\int_0^T v dt, \quad (2.192)$$

respectively. In POD, we wish to express the data, v, as

$$v(x,t) = \sum_i^N f_i(t)u_i(x), \quad (2.193)$$

where both, $f_i(t)$ and the spatial basis functions $u_i(x)$, have to be found. We assume all the functions involved are continuous, and the integral over space and the integral over time commute. To start the process, let us ask ourselves the following: What is the single term $f(t)u(x)$, which is the best approximation for $v(x,t)$?

The criterion for finding u is the requirement that the expectation value of the square-error E has to be a minimum. The square-error E is defined as

$$E = \int_\Omega [v(x,t) - f(t)u(x)]^2 dx, \quad (2.194)$$

and its expectation value, as

$$\mathbb{E}\{E\} = \frac{1}{T}\int_0^T \int_\Omega [v(x,t) - f(t)u(x)]^2 dx \, dt. \quad (2.195)$$

We assume our single basis function is normalized. That is,

$$\|u\| = 1. \quad (2.196)$$

The extremization of $\mathbb{E}\{E\}$ with the constraint $\|u\| = 1$ can be done using the modified functional

$$\mathbb{E}\{E^*\} = \mathbb{E}\{E\} + \mu[\|u\|^2 - 1], \quad (2.197)$$

where μ is a Lagrange multiplier. Explicitly,

$$\mathbb{E}\{E^*\} = \frac{1}{T}\int_0^T \int_\Omega [v(x,t) - f(t)u(x)]^2 dx \, dt + \mu[\|u\|^2 - 1]. \quad (2.198)$$

The first variation of this functional using $f \longrightarrow f + \delta f$ and $u \longrightarrow u + \delta u$, gives

$$\frac{1}{T} \int_0^T \int_\Omega 2[v - fu](-u\delta f)dx\, dt = 0, \tag{2.199}$$

$$\frac{1}{T} \int_0^T \int_\Omega 2[v - fu](-f\delta u)dx\, dt + 2 \int_\Omega u\delta u dx = 0. \tag{2.200}$$

Noting that f is a function of only t, and u is that of only x, we have

$$f = \langle v, u \rangle, \tag{2.201}$$

$$\frac{1}{T} \int_0^T [vf - f^2 u]dt - \mu u = 0, \tag{2.202}$$

where we have used $\|u\| = 1$. Using Eq. (2.201) in (2.202),

$$\int_\Omega \mathbb{E}\{v(x,t)v(x',t)\}u(x')dx' = [\mathbb{E}\{f^2\} + \mu]u(x). \tag{2.203}$$

This relation shows that the function u gets mapped into itself through a symmetric integral operator $R(x,x')$ defined by

$$R(x,x') = \mathbb{E}\{v(x,t)v(x',t)\}, \tag{2.204}$$

and the associated eigenvalue problem is

$$\int_\Omega R(x,x')u(x')dx' = \frac{1}{\lambda}u(x). \tag{2.205}$$

Equation (2.203) becomes

$$\frac{1}{\lambda} = \mathbb{E}\{f^2\} + \mu. \tag{2.206}$$

Let us compute the first term on the right-hand side.

$$
\begin{aligned}
\mathbb{E}\{f^2\} &= \mathbb{E}\{\int_\Omega v(x,t)u(x)dx\}^2 \\
&= \mathbb{E}\{\int_\Omega v(x,t)u(x)dx \int_\Omega v(x',t)u(x')dx'\} \\
&= \int_\Omega \int_\Omega \mathbb{E}\{v(x,t)v(x',t)\}u(x')dx'u(x)dx \\
&= \int_\Omega \frac{u^2(x)}{\lambda}dx \\
&= \frac{1}{\lambda}.
\end{aligned}
\tag{2.207}
$$

Thus, the Lagrange multiplier μ turns out to be zero. Also, as $\mathbb{E}\{f^2\} > 0$, the eigenvalue is positive.

From the Hilbert-Schmidt theory, our eigenvalue problem with a real symmetric continuous kernel gives a set of N orthonormal eigenfunctions u_i with eigenvalues λ_i, where N may be infinite. With this, we extend our approximation to

$$
v(x,t) = \sum_i^N f_i(t)u_i(x),
\tag{2.208}
$$

where $f_i = \langle v, u_i \rangle$ and the orthonormal eigenfunctions u_i satisfy

$$
u_i(x) = \lambda_i \int_\Omega R(x,x')u_i(x')dx'.
\tag{2.209}
$$

The expectation value of the error becomes

$$
\mathbb{E}\{E\} = \mathbb{E}\{\int_\Omega [v - \sum f_i u_i]^2 dx\},
\tag{2.210}
$$

which may be broken into the three integrals:

$$I_1 = \mathbb{E}\{\int_\Omega v(x,t)v(x,t)dx\},$$

$$I_2 = -2\sum \mathbb{E}\{\int_\Omega v(x,t)f_i u_i(x)dx\},$$

$$I_3 = \sum \mathbb{E}\{\int_\Omega f_i^2 u_i^2(x)dx\},$$

where in I_3, we have anticipated the orthogonality of u_i.

With the expansion of the kernel using Mercer's theorem stated in Eq. (2.100), we have

$$\mathbb{E}\{v(x,t)v(x',t)\} = R(x,x') = \sum \frac{u_i(x)u_i(x')}{\lambda_i},$$

$$I_1 = \int_\Omega R(x,x)dx = \sum \frac{1}{\lambda_i},$$

$$I_2 = -2\sum \mathbb{E}\{f_i^2\} = -2\sum \frac{1}{\lambda_i},$$

$$I_3 = \sum \mathbb{E}\{f_i^2\} = \sum \frac{1}{\lambda_i}. \tag{2.211}$$

These integrals add up to zero, and our choice of basis functions makes the expectation value of the error zero, provided the complete sequence of eigenfunctions are used for the expansion.

In applications, an approximate representation of the data using a finite number of eigenfunctions is more practical. Also, in order to improve numerical accuracy the spatial mean value of the data is removed before finding the low dimensional representation through the proper orthogonal decomposition.

SUGGESTED READING

Abramowitz, M. and Stegun, I. (1965). *Handbook of Mathematical Functions* (National Bureau of Standards), Dover.

Barber, J. R. (2002). *Elasticity*, 2nd ed., Kluwer.

Brebbia, C. A. (1978). *The Boundary Element Method for Engineers*, Pentech Press, London.

Chatterjee, A. (2000). An introduction to the proper orthogonal decomposition, *Current Science*, Vol. 78, No. 7, pp. 808–817.

Courant, R. and Hilbert, D. (1953). *Methods of Mathematical Physics*, Vol. 2, Interscience.

Hartmann, F. (1989). *Introduction to Boundary Elements*, Springer-Verlag.

Hildebrand, F. B. (1965). *Methods of Applied Mathematics*, 2nd ed., Prentice-Hall.

Holmes, P., Lumley, J. L., and Berkooz, G. (1996). *Turbulence, Coherent Structures, Dynamical Systems and Symmetry*, Cambridge University Press.

Pozrikides, C. (2002). *A Practical Guide to Boundary Element Methods*, Chapman & Hall/CRC.

Rahman, M. (2007). *Integral Equations and Their Applications*, WIT Press.

Sneddon, I. N. (1966). *Mixed Boundary-Value Problems in Potential Theory*, North Holland.

Tricomi, F. G. (1957). *Integral Equations*, Interscience.

EXERCISES

2.1 Using differentiation, convert the integral equation

$$u(x) = \frac{1}{2} \int_0^1 |x - \xi| u(\xi) d\xi + \frac{x^2}{2}$$

into a differential equation. Obtain the needed boundary conditions, and solve the differential equation.

2.2 Convert

$$u(x) = \lambda \int_{-1}^1 e^{-|x-\xi|} u(\xi) d\xi$$

into a differential equation. Obtain the required number of boundary conditions.

2.3 Solve the integral equation

$$u(x) = e^x + \lambda \int_{-1}^1 e^{(x-\xi)} u(\xi) d\xi,$$

and discuss the conditions on λ for a unique solution.

2.4 Solve

$$u(x) = x + \int_0^1 (1 - x\xi) u(\xi) d\xi.$$

2.5 Solve

$$u(x) = \sin x + \int_0^\pi \cos(x - \xi)u(\xi)d\xi.$$

2.6 Obtain the eigenvalues and eigenfunctions of the equation

$$u(x) = \lambda \int_0^{2\pi} \cos(x - \xi)u(\xi)d\xi.$$

2.7 Find the eigenvalues and eigenfunctions of

$$u(x) = \lambda \int_0^1 (1 - x\xi)u(\xi)d\xi.$$

2.8 From the differential equation and the boundary conditions obtained for Exercise 2.2, find the values of λ for the existence of non-trivial solutions.

2.9 Show that the equation

$$u(x) = \lambda \int_0^1 |x - \xi|u(\xi)d\xi$$

has an infinite number of eigenvalues and eigenfunctions.

2.10 Obtain the values of λ and a for the equation

$$u(x) = \lambda \int_0^1 (x - \xi)u(\xi)d\xi + a + x^2$$

to have (a) a unique solution, (b) a nonunique solution, and (c) no solution.

2.11 Obtain the resolvent kernel for

$$u(x) = \lambda \int_0^{2\pi} e^{in(x-\xi)}u(\xi)d\xi + f(x).$$

2.12 Find the resolvent kernel for

$$u(x) = \lambda \int_0^\pi \cos(x - \xi)u(\xi)d\xi + f(x).$$

2.13 Show that for a Fredholm equation

$$u(x) = \lambda \int_a^b k(x,\xi)u(\xi)d\xi + f(x),$$

the resolvent kernel $g(x,\xi)$ satisfies

$$g(x,\xi) = k(x,\xi) + \lambda \int_a^b k(x,\eta)g(\eta,\xi)d\eta.$$

Thus, the resolvent kernel satisfies the integral equation for u when the forcing function f is replaced by k with ξ and λ kept as parameters.

2.14 Demonstrate the preceding result when $k = 1 - 2x\xi$ in the domain $0 < x,\xi < 1$.

2.15 For the Volterra equation,

$$u(x) = \int_0^x (1 - x\xi)u(\xi)d\xi + 1,$$

starting with $u^{(0)} = 1$, obtain iteratively, $u^{(2)}$.

2.16 For the integral equation of the first kind

$$\int_0^{2\pi} \sin(x + \xi)u(\xi)d\xi = \sin x,$$

discuss the consequence of assuming

$$u(x) = \sum_{n=0} a_n \cos nx + \sum_{n=1} b_n \sin nx,$$

where the constants, a_n and b_n, are unknown.

2.17 With the quadratic forms,

$$J_1[u] = \int_a^b \int_a^b k(x,\xi)u(x)u(\xi)d\xi\,dx, \quad J_2[u] = \int_a^b u^2 dx,$$

where u is a smooth function in (a,b), show that the functions u, which extremize

$$J[u] = \frac{J_1[u]}{J_2[u]},$$

are the eigenfunctions of $[k(x,\xi) + k(\xi,x)]/2$. Also show that the extrema of J correspond to the reciprocal of the eigenvalues of $k(x,\xi)$.

2.18 In the preceding problem, assuming k is symmetric, compute $J[v]$ if

$$v(x) = \sum_n a_n u_n(x),$$

where u_n are normalized eigenfunctions.

2.19 The generalized Abel equation,

$$\int_0^x \frac{u(\xi)d\xi}{(x-\xi)^\alpha} = f(x),$$

is solved by multiplying both sides by $(\eta - x)^{\alpha-1}$ and integrating with respect to x from 0 to η. Implement this procedure, and discuss the allowable range for the index, α.

2.20 Solve the singular equation

$$\int_0^x \frac{u(\xi)d\xi}{(x^2-\xi^2)} = f(x),$$

using a change of the independent variable.

2.21 If an elastic, 3D half space ($z > 0$) is subjected to an axi-symmetric z-displacement, $w(r)$, where $r = \sqrt{x^2+y^2}$, we need to solve the integral equation for the distributed pressure on the horizontal surface, $z = 0$,

$$\int_0^r \frac{g(\rho)d\rho}{\sqrt{r^2-\rho^2}} = \frac{\mu}{1-\nu}w(r), \quad 0 < r < a,$$

where μ is the shear modulus, ν is the Poisson's ratio, and a is the contact radius. Obtain the pressure distribution $g(r)$ (see Barber (2002)).

2.22 In the preceding problem, if the contact displacement is due to a rigid sphere of radius R pressing against the elastic half space, we assume

$$w(r) = d - \frac{r^2}{2R},$$

where d is the maximum indentation. Obtain the pressure distribution, the value of the contact radius a, and the total vertical force.

2.23 Consider the integral equation

$$u(x) = \int_0^1 k(x,\xi)u(\xi)d\xi + x^2,$$

where

$$k(x,\xi) = \begin{cases} x(\xi-1), & x < \xi, \\ \xi(x-1), & x > \xi. \end{cases}$$

Obtain an exact solution to this equation. Using the approximate kernel $Ax(1-x)$, find A using the method of least square error. Obtain an approximate solution for u. Compare the values of the exact and approximate solutions at $x = 0.5$.

2.24 Consider the differential equation

$$u'' + u = x, \quad u(0) = 0, \quad u(1) = 0.$$

Find an exact solution. Obtain a finite difference solution by dividing the domain into four equal intervals.

Convert the differential equation into an integral equation using the Green's function for the operator $L = d^2/dx^2$. Obtain a numerical solution of the integral equation by dividing the domain into four intervals and using the trapezoidal rule and the collocation method. Compare the results with the exact solution.

2.25 Consider the nonlinear integral equation

$$u(x) - \lambda \int_0^1 u^2(\xi)d\xi = 1.$$

Show that for $\lambda < 1/4$ this equation has two solutions and for $\lambda > 1/4$ there are no real solutions. Also show that one of these solutions is singular at $\lambda = 0$ and two solutions coalesce at $\lambda = 1/4$. Sketch the solutions as functions of λ. (Based on Tricomi (1957).)

3

FOURIER TRANSFORMS

The method of Fourier transforms is a powerful technique for solving linear, partial differential equations arising in engineering and physics when the domain is infinite or semi-infinite. This is an extension of the Fourier series, which is applicable to periodic functions defined on an interval $-a < x < a$. First, let us review the Fourier series and extend it to infinite domains in a heuristic form.

3.1 FOURIER SERIES

If $f(x)$ is a continuous function on $-a < x < a$, we expand it in terms of the orthogonal functions,

$$\{1, \cos(\pi x/a), \cos(2\pi x/a), \ldots, \sin(\pi x/a), \sin(2\pi x/a), \ldots\}$$

as

$$f(x) = A_0 + \sum_{n=1}^{\infty} [A_n \cos(n\pi x/a) + B_n \sin(n\pi x/a)]. \tag{3.1}$$

Taking the inner product of this equation with each member of the orthogonal set (basis), we obtain

$$A_0 = \frac{1}{2a} \int_{-a}^{a} f(t)dt, \tag{3.2}$$

$$A_n = \frac{1}{a} \int_{-a}^{a} f(t) \cos(n\pi t/a) \, dt, \tag{3.3}$$

$$B_n = \frac{1}{a} \int_{-a}^{a} f(t) \sin(n\pi t/a) \, dt. \tag{3.4}$$

As the functions in our orthogonal basis are periodic, if $f(x)$ is evaluated for a value of x outside the domain $-a < x < a$, we find a periodic extension of f.

We may also form the complex form of the Fourier series using the basis

$$\{e^{-i\pi nx/a}, \quad -\infty < n < \infty\}.$$

Here, the negative sign for the exponent is chosen to have compatibility with our convention for Fourier transforms. Writing

$$f(x) = \sum_{n=-\infty}^{\infty} C_n e^{-i\pi nx/a}, \tag{3.5}$$

we find

$$C_n = \frac{1}{2a} \int_{-a}^{a} f(t) e^{i\pi nt/a}\, dt, \tag{3.6}$$

where the orthogonality of the basis,

$$\int_{-a}^{a} e^{i(m-n)\pi x/a}\, dx = 2a\delta_{mn}, \tag{3.7}$$

was used.

Substituting for C_n in the complex Fourier series representation, we get the identity

$$f(x) = \frac{1}{2a} \sum_{n=-\infty}^{\infty} e^{-i\pi nx/a} \int_{-a}^{a} f(t) e^{i\pi nt/a}\, dt. \tag{3.8}$$

This equations applies only when f is continuous.

3.2 FOURIER TRANSFORM

Historically, the Fourier transform was introduced by extending the domain to infinity by taking the limit, $a \to \infty$. This limiting process needs to be carried out carefully as, in the exponent, $in\pi x/a$, n may also be large. Let

$$\xi = \frac{n\pi}{a}, \quad \Delta\xi = \frac{\pi}{a}.$$

Then the preceding identity, Eq. (3.8), becomes

$$f(x) = \frac{1}{2\pi} \sum e^{-i\xi x} \int_{-a}^{a} f(t)e^{i\xi t} \, dt \Delta \xi. \tag{3.9}$$

As $a \to \infty$, we get the Fourier integral theorem

$$f(x) = \frac{1}{2\pi} \int_{-\infty}^{\infty} e^{-i\xi x} \int_{-\infty}^{\infty} f(t)e^{i\xi t} \, dt \, d\xi. \tag{3.10}$$

We may express the double integral on the right side as a single integral, using

$$F(\xi) = \frac{1}{\sqrt{2\pi}} \int_{-\infty}^{\infty} f(x)e^{i\xi x} \, dx, \tag{3.11}$$

$$f(x) = \frac{1}{\sqrt{2\pi}} \int_{-\infty}^{\infty} F(\xi)e^{-i\xi x} \, d\xi. \tag{3.12}$$

These two equations define the Fourier transform $F(\xi)$ and its inverse transform $f(x)$, when $f(x)$ is continuous. We use upper case letters for the transform in the ξ domain and lower case for the inverse in the x domain.

We may introduce integral operators:

$$\mathcal{F}[f] = \int_{-\infty}^{\infty} e^{ix\xi} f(x) \, dx, \tag{3.13}$$

$$\mathcal{F}^{-1}[F] = \int_{-\infty}^{\infty} e^{-ix\xi} F(\xi) \, d\xi. \tag{3.14}$$

As the kernels of these integrals are complex conjugates of each other, we have

$$\mathcal{F}^{-1} = \mathcal{F}^*, \quad \mathcal{F}\mathcal{F}^* = \mathcal{I}, \tag{3.15}$$

where an asterisk denotes the complex conjugate and $\mathcal{I}$ is the identity operator.

We have introduced $1/\sqrt{2\pi}$ in both the transform and its inverse, which gives a convenient symmetry to the representations. Electrical engineers remove this factor from the transform and use $1/(2\pi)$ in the inverse. They also use $(-i)$ in the transform and (i) in the inverse. Also, electrical engineers use j for the imaginary number to avoid confusion with the notation i for current. These conventions do not alter the Fourier integral theorem.

Our informal approach to the Fourier integral theorem can be put in sharper focus by clearly defining the class of functions amenable for this transformation. We need f to be absolutely integrable and piecewise continuous.

A function is absolutely integrable on $\mathbf{R}$ (i.e., $f \in \mathcal{A}$), if

$$\int_{-\infty}^{\infty} |f(x)| \, dx < M, \tag{3.16}$$

where M is finite. This can also be stated as: Given a number ϵ, there exist numbers R_1 and R_2 such that

$$\int_{R_1}^{\infty} |f(x)| \, dx \leq \epsilon, \quad \int_{-\infty}^{R_2} |f(x)| \, dx \leq \epsilon, \tag{3.17}$$

A function is piecewise continuous on $\mathbf{R}$ (i.e., $f \in \mathcal{P}$), if

$$\lim_{\epsilon \to 0} [f(x+\epsilon) - f(x-\epsilon)] = f(x^+) - f(x^-) \neq 0, \tag{3.18}$$

for a finite number of values of x.

When the functions of the class $\mathcal{P}$ are included, the Fourier integral theorem reads:

$$\frac{1}{2}[f(x^+) + f(x^-)] = \frac{1}{2\pi} \int_{-\infty}^{\infty} e^{-i\xi x} \int_{-\infty}^{\infty} f(t) e^{i\xi t} \, dt \, d\xi. \tag{3.19}$$

Thus, the inverse transform gives the mean value of the left and the right limits of the function at any x.

Before we attempt to prove this, it is helpful to establish two results: the Riemann-Lebesgue lemma and the localization lemma.

3.2.1 Riemann-Lebesgue Lemma

If $f(x)$ is a continuous function in (a,b),

$$I_s \equiv \lim_{\lambda \to \infty} \int_a^b f(x) \sin \lambda x \, dx = 0, \qquad (3.20)$$

$$I_c \equiv \lim_{\lambda \to \infty} \int_a^b f(x) \cos \lambda x \, dx = 0. \qquad (3.21)$$

As shown in Fig. 3.1, multiplication by $\sin \lambda x$ or $\cos \lambda x$ slices the function and creates areas of alternating signs under the curve. To prove this lemma, we use the periodicity of the trigonometric functions to write

$$\sin \lambda(x + \pi/\lambda) = -\sin \lambda x. \qquad (3.22)$$

$$I_s = \int_a^b f(x) \sin \lambda x \, dx = -\int_a^b f(x) \sin \lambda(x + \pi/\lambda) \, dx. \qquad (3.23)$$

$$2I_s = \int_a^b f(x) \sin \lambda x \, dx - \int_a^b f(x) \sin \lambda(x + \pi/\lambda) \, dx. \qquad (3.24)$$

Replacing $x + \pi/\lambda$ by x' in the second integral,

$$2I_s = \int_a^b f(x) \sin \lambda x \, dx - \int_{a-\pi/\lambda}^{b-\pi/\lambda} f(x - \pi/\lambda) \sin \lambda x \, dx, \qquad (3.25)$$

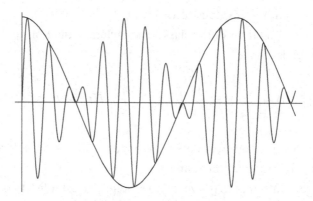

Figure 3.1. Product of $f(x) = \cos x$ and $\sin \lambda x$ ($\lambda = 10$).

where we have again replaced x' by x. The integrals on the right-hand side can be written as

$$2I_s = -\int_{a-\pi/\lambda}^{a} f(x-\pi/\lambda)\sin\lambda x\,dx$$

$$+ \int_{a}^{b} [f(x)-f(x-\pi/\lambda)]\sin\lambda x\,dx \qquad (3.26)$$

$$+ \int_{b-\pi/\lambda}^{b} f(x-\pi/\lambda)\sin\lambda x\,dx.$$

As $\lambda \to \infty$, we see that each one of the three integrals goes to zero. The integral I_c can be shown to vanish in a similar way.

3.2.2 Localization Lemma

For a continuous function $f(x)$ defined in $(0,a)$,

$$I_0 \equiv \lim_{\lambda\to\infty} \int_0^a f(x)\frac{\sin\lambda x}{x}\,dx = \frac{\pi}{2}f(0^+). \qquad (3.27)$$

To see this result, we write I_0 as

$$I_0 = \lim_{\lambda\to\infty}\left\{ \int_0^a [f(x)-f(0^+)]\frac{\sin\lambda x}{x}\,dx + f(0^+)\int_0^a \frac{\sin\lambda x}{x}\,dx\right\}. \qquad (3.28)$$

By the Riemann-Lebesgue lemma, the first integral goes to zero as $[f(x)-f(0^+)]/x$ is continuous. In the second integral, substituting $\lambda x = y$,

$$\int_0^{a\lambda} \frac{\sin y}{y}\,dy. \qquad (3.29)$$

As $\lambda \to \infty$, we have

$$\int_0^\infty \frac{\sin x}{x}\,dx = \frac{\pi}{2}, \qquad (3.30)$$

where the last integral can be obtained by integrating the complex function e^{iz}/z along the real line and using the residue theorem. Thus, we have the localization lemma stated in Eq. (3.27).

Also, note that the sequence

$$\delta_\lambda(x) = \frac{1}{\pi}\frac{\sin\lambda x}{x} \qquad (3.31)$$

forms a δ-sequence, which converges to the Dirac delta function as $\lambda \to \infty$.

3.3 FOURIER INTEGRAL THEOREM

We begin by considering the integral on the right-hand side of Eq. (3.19),

$$I = \frac{1}{2\pi} \int_{-\infty}^{\infty} e^{-i\xi x} \int_{-\infty}^{\infty} f(t)e^{i\xi t}\, dt\, d\xi$$

$$= \frac{1}{2\pi} \int_{-\infty}^{\infty} f(t)\left[\int_0^{\infty} e^{i\xi(t-x)}\, d\xi + \int_{-\infty}^0 e^{i\xi(t-x)}\, d\xi\right] dt. \quad (3.32)$$

In the second integral inside the brackets, we let $\xi \to -\xi$ and the new upper limits of ∞ in both the integrals is replaced by λ, which would go to infinity in the limit. Then

$$I = \lim_{\lambda\to\infty} \frac{1}{2\pi} \int_{-\infty}^{\infty} f(t) \int_0^{\lambda} [e^{i\xi(t-x)} + e^{-i(t-x)}]\, d\xi\, dt$$

$$= \lim_{\lambda\to\infty} \frac{1}{\pi} \int_{-\infty}^{\infty} f(t) \int_0^{\lambda} \cos\xi(t-x)d\xi\, dt$$

$$= \lim_{\lambda\to\infty} \frac{1}{\pi} \int_{-\infty}^{\infty} f(t)\frac{\sin\lambda(t-x)}{t-x}\, dt$$

$$= \lim_{\lambda\to\infty} \frac{1}{\pi} \int_{-\infty}^{\infty} f(\tau+x)\frac{\sin\lambda\tau}{\tau}\, d\tau$$

$$= \lim_{\lambda\to\infty} \frac{1}{\pi} \left[\int_0^{\infty} f(\tau+x)\frac{\sin\lambda\tau}{\tau}\, d\tau + \int_{-\infty}^0 f(\tau+x)\frac{\sin\lambda\tau}{\tau}\, d\tau\right]$$

$$= \frac{1}{2}[f(x^+)+f(x^-)], \quad (3.33)$$

where we have used the localization lemma in the last step.

Having established the Fourier integral theorem for functions of class $\mathcal{P}$, hereafter we confine our attention to functions of class (continuous functions) for simplicity. We have two domains: the x domain where functions f are given and the ξ domain where the transforms F exist.

3.4 FOURIER COSINE AND SINE TRANSFORMS

When a function $f(x)$ is given on $(0, \infty)$, we may extend it to $(-\infty, 0)$ as an even or an odd function. Figure 3.2 shows the extended functions,

$$f_e = \begin{cases} f(x), & x > 0, \\ f(-x), & x < 0, \end{cases} \tag{3.34}$$

$$f_o = \begin{cases} f(x), & x > 0, \\ -f(-x), & x < 0, \end{cases} \tag{3.35}$$

We can take the Fourier transform of these extended functions provided they are absolutely integrable and piecewise continuous.

$$\mathcal{F}[f_e] = \frac{1}{\sqrt{2\pi}} \int_{-\infty}^{\infty} f_e(x) e^{i\xi x}\, dx$$

$$= \frac{1}{\sqrt{2\pi}} \left\{ \int_{0}^{\infty} f_e(x) e^{i\xi x}\, dx + \int_{-\infty}^{0} f_e(x) e^{i\xi x}\, dx \right\}$$

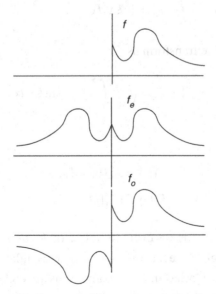

Figure 3.2. Even and odd extensions of a function $f(x)$.

$$= \frac{1}{\sqrt{2\pi}} \left\{ \int_0^\infty f_e(x)e^{i\xi x}\,dx + \int_0^\infty f_e(-x)e^{-i\xi x}\,dx \right\}$$

$$= \sqrt{\frac{2}{\pi}} \int_0^\infty f(x)\cos\xi x\,dx. \tag{3.36}$$

We define the Cosine transform as

$$\mathcal{F}_c[f] = F_c = \sqrt{\frac{2}{\pi}} \int_0^\infty f(x)\cos\xi x\,dx. \tag{3.37}$$

Similarly,

$$\mathcal{F}[f_o] = \frac{1}{\sqrt{2\pi}} \int_{-\infty}^\infty f_o(x)e^{i\xi x}\,dx$$

$$= \frac{1}{\sqrt{2\pi}} \left\{ \int_0^\infty f_o(x)e^{i\xi x}\,dx + \int_{-\infty}^0 f_o(x)e^{i\xi x}\,dx \right\}$$

$$= \frac{1}{\sqrt{2\pi}} \left\{ \int_0^\infty f_o(x)e^{i\xi x}\,dx + \int_0^\infty f_o(-x)e^{-i\xi x}\,dx \right\}$$

$$= i\sqrt{\frac{2}{\pi}} \int_0^\infty f(x)\sin\xi x\,dx. \tag{3.38}$$

We define the Sine transform as

$$\mathcal{F}_s[f] = F_s = \sqrt{\frac{2}{\pi}} \int_0^\infty f(x)\sin\xi x\,dx. \tag{3.39}$$

Then

$$\mathcal{F}[f_e] = \mathcal{F}_c[f] = F_c, \tag{3.40}$$

$$\mathcal{F}[f_o] = i\mathcal{F}_s[f] = iF_s. \tag{3.41}$$

We note that $F_c(\xi)$ is an even function of ξ, as ξ appears in the definition through the even function $\cos\xi x$. Similarly, F_s is an odd function of ξ. The Cosine and Sine transforms are collectively referred to as **Trigonometric transforms**. The inversion formulas for the trigonometric transforms may be developed as follows:

$$f_e(x) = \frac{1}{\sqrt{2\pi}} \int_{-\infty}^{\infty} F_c(\xi) e^{-i\xi x} \, d\xi$$

$$= \sqrt{\frac{2}{\pi}} \int_0^{\infty} F_c(\xi) \cos \xi x \, d\xi,$$

$$f(x) = \sqrt{\frac{2}{\pi}} \int_0^{\infty} F_c(\xi) \cos \xi x \, d\xi, \quad x > 0, \tag{3.42}$$

where we have used the evenness property of the function $\cos \xi x$. Similarly,

$$f_o(x) = i \sqrt{\frac{2}{\pi}} \int_{-\infty}^{\infty} F_s(\xi) e^{-i\xi x} \, d\xi$$

$$= \sqrt{\frac{2}{\pi}} \int_0^{\infty} F_s(\xi) \sin \xi x \, d\xi,$$

$$f(x) = \sqrt{\frac{2}{\pi}} \int_0^{\infty} F_s(\xi) \sin \xi x \, d\xi, \quad x > 0, \tag{3.43}$$

where we have used the oddness property of the function $\sin \xi x$. The trigonometric transform operators have the properties

$$\mathcal{F}_c^{-1} = \mathcal{F}_c, \quad \mathcal{F}_s^{-1} = \mathcal{F}_s. \tag{3.44}$$

If $f(x)$ is defined on $(-\infty, \infty)$, we may obtain an even and odd component by introducing

$$f_e(x) = \frac{1}{2}[f(x) + f(-x)], \quad f_o = \frac{1}{2}[f(x) - f(-x)]. \tag{3.45}$$

$$f(x) = f_e(x) + f_o(x). \tag{3.46}$$

This decomposition scheme, when applied to e^x, gives the even function $\cosh x$ and the odd function $\sinh x$.

With this decomposition,

$$\mathcal{F}[f] = \mathcal{F}[f_e + f_o] = F_c + iF_s, \tag{3.47}$$

$$\mathcal{F}^* \mathcal{F}[f] = [\mathcal{F}_c - i\mathcal{F}_s][\mathcal{F}_c + i\mathcal{F}_s] = f_e + f_o = f. \tag{3.48}$$

If the given function f is even,

$$F = F_c, \quad f = [\mathcal{F}_c - i\mathcal{F}_s][F_c] = \mathcal{F}_c[F_c], \tag{3.49}$$

and if it is odd,

$$F = iF_s, \quad f = [\mathcal{F}_c - i\mathcal{F}_s][iF_s] = \mathcal{F}_s[F_s]. \tag{3.50}$$

3.5 PROPERTIES OF FOURIER TRANSFORMS

From the definition

$$F(\xi) = \frac{1}{\sqrt{2\pi}} \int_{-\infty}^{\infty} f(x)e^{i\xi x}\,dx, \tag{3.51}$$

using the Riemann-Lebesgue lemma, we see

$$\lim_{|\xi| \to \infty} F(\xi) = 0. \tag{3.52}$$

We can also see that $F(\xi)$ is continuous, because

$$F(\xi + h) - F(\xi) = \frac{1}{\sqrt{2\pi}} \int_{-\infty}^{\infty} f(x)e^{i\xi x}\left[e^{ihx} - 1\right]dx \tag{3.53}$$

goes to zero as $h \to 0$.

3.5.1 Derivatives of F

Differentiating the defining relation,

$$F'(\xi) = \frac{i}{\sqrt{2\pi}} \int_{-\infty}^{\infty} xf(x)e^{i\xi x}\,dx = \mathcal{F}[ixf], \tag{3.54}$$

which exists if $xf(x) \in \mathcal{A}$. Differentiating n times,

$$F^{(n)}(\xi) = \mathcal{F}[(ix)^n f], \tag{3.55}$$

if $x^n f \in \mathcal{A}$.

3.5.2 Scaling

$$\mathcal{F}[f(x/a)] = \frac{1}{\sqrt{2\pi}} \int_{-\infty}^{\infty} f(x/a)e^{i\xi x}\, dx$$

$$= \frac{a}{\sqrt{2\pi}} \int_{-\infty}^{\infty} f(x)e^{ia\xi x}\, dx = aF(a\xi), \qquad (3.56)$$

where we have assumed the constant $a > 0$.

3.5.3 Phase Change

$$\mathcal{F}[f(x)e^{iax}] = \frac{1}{\sqrt{2\pi}} \int_{-\infty}^{\infty} f(x)e^{i(\xi+a)x}\, dx = F(\xi + a). \qquad (3.57)$$

3.5.4 Shift

$$\mathcal{F}[f(x-a)] = \frac{1}{\sqrt{2\pi}} \int_{-\infty}^{\infty} f(x-a)e^{i\xi x}\, dx = e^{ia\xi}F(\xi). \qquad (3.58)$$

3.5.5 Derivatives of f

$$\mathcal{F}[f'(x)] = \frac{1}{\sqrt{2\pi}} \int_{-\infty}^{\infty} f'(x)e^{i\xi x}\, dx$$

$$= \frac{1}{\sqrt{2\pi}} \left[f(x)e^{i\xi x} \Big|_{-\infty}^{\infty} - i\xi \int_{-\infty}^{\infty} f(x)e^{i\xi x}\, dx \right]$$

$$= (-i\xi)F(\xi), \qquad (3.59)$$

where we used $f = 0$ at the limits of integration. This result shows the most useful property of the Fourier transform: The differential operator in the x domain becomes an algebraic operator in the ξ domain.

Differentiating further,

$$\mathcal{F}[f^{(n)}(x)] = \frac{1}{\sqrt{2\pi}} \int_{-\infty}^{\infty} f^{(n)}(x)e^{i\xi x}\, dx$$

$$= \frac{1}{\sqrt{2\pi}} (-i\xi)^n \int_{-\infty}^{\infty} f(x)e^{i\xi x}\, dx$$

$$= (-i\xi)^n F(\xi). \qquad (3.60)$$

3.6 PROPERTIES OF TRIGONOMETRIC TRANSFORMS

With the definitions

$$F_c = \sqrt{\frac{2}{\pi}} \int_0^\infty f(x) \cos \xi x \, dx, \quad F_s = \sqrt{\frac{2}{\pi}} \int_0^\infty f(x) \sin \xi x \, dx, \quad (3.61)$$

we list some of the useful properties of the trigonometric transforms next.

3.6.1 Derivatives of F_c and F_s

$$F_c' = -\mathcal{F}_s[xf], \quad F_s' = \mathcal{F}_c[xf]. \quad (3.62)$$

3.6.2 Scaling

For positive numbers, a, the scaling operation gives

$$\mathcal{F}_c[f(x/a)] = aF_c(a\xi), \quad \mathcal{F}_s[f(x/a)] = aF_s(a\xi). \quad (3.63)$$

3.6.3 Derivatives of f

Using integration by parts, we see

$$\mathcal{F}_c[f'(x)] = \sqrt{\frac{2}{\pi}} \int_0^\infty f'(x) \cos \xi x \, dx$$

$$= -\sqrt{\frac{2}{\pi}} f(0) + \xi F_s(\xi), \quad (3.64)$$

$$\mathcal{F}_s[f'(x)] = -\xi F_c(\xi), \quad (3.65)$$

$$\mathcal{F}_c[f''(x)] = -\sqrt{\frac{2}{\pi}} f'(0) + \xi \mathcal{F}_s[f']$$

$$= -\sqrt{\frac{2}{\pi}} f'(0) - \xi^2 F_c(\xi), \quad (3.66)$$

$$\mathcal{F}_s[f''(x)] = -\xi \mathcal{F}_c[f']$$

$$= \sqrt{\frac{2}{\pi}} \xi f(0) - \xi^2 F_s(\xi). \quad (3.67)$$

The presence of the boundary terms in the transforms of the second derivatives directs one to choose the cosine transform if $f'(0)$ is known and the sine transform if $f(0)$ is known.

3.7 EXAMPLES: TRANSFORMS OF ELEMENTARY FUNCTIONS

Now we are ready to obtain the transforms of some familiar functions belonging to the class $\mathcal{A}$.

3.7.1 Exponential Functions

Let us begin with the integrals

$$I = \int_0^\infty e^{-ax} \cos \xi x \, dx, \quad J = \int_0^\infty e^{-ax} \sin \xi x \, dx. \quad (3.68)$$

Using integration by parts,

$$I = \frac{e^{-ax}}{-a} \cos \xi x \Big|_0^\infty - \frac{\xi}{a} \int_0^\infty e^{-ax} \sin \xi x \, dx = \frac{1}{a} - \frac{\xi}{a} J,$$

$$J = \frac{e^{-ax}}{-a} \sin \xi x \Big|_0^\infty + \frac{\xi}{a} \int_0^\infty e^{-at} \cos \xi x \, dx = \frac{\xi}{a} I. \quad (3.69)$$

Then

$$I = \frac{a}{a^2 + \xi^2}, \quad J = \frac{\xi}{a^2 + \xi^2}. \quad (3.70)$$

These results and the property that the trigonometric transform operators and their inverses are identical lead to

$$\mathcal{F}_c[e^{-ax}] = \sqrt{\frac{2}{\pi}} \frac{a}{a^2 + \xi^2}, \quad \mathcal{F}_c\left[\sqrt{\frac{2}{\pi}} \frac{a}{a^2 + x^2}\right] = e^{-a\xi}, \quad (3.71)$$

$$\mathcal{F}_s[e^{-ax}] = \sqrt{\frac{2}{\pi}} \frac{\xi}{a^2 + \xi^2}, \quad \mathcal{F}_s\left[\sqrt{\frac{2}{\pi}} \frac{x}{a^2 + x^2}\right] = e^{-a\xi}. \quad (3.72)$$

Thus, every time we compute a transform, we get another one for free!

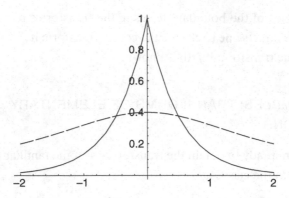

Figure 3.3. The function $e^{-2|x|}$ and its Fourier transform (dashed curve).

Using the even and odd extensions of this function, we obtain

$$\mathcal{F}[e^{-a|x|}] = \sqrt{\frac{2}{\pi}}\frac{a}{a^2+\xi^2}, \quad \mathcal{F}\left[\sqrt{\frac{2}{\pi}}\frac{a}{a^2+x^2}\right] = e^{-a|\xi|}, \qquad (3.73)$$

$$\mathcal{F}[\text{sgn}(x)e^{-a|x|}] = i\sqrt{\frac{2}{\pi}}\frac{\xi}{a^2+\xi^2}, \quad \mathcal{F}\left[\sqrt{\frac{2}{\pi}}\frac{x}{a^2+x^2}\right] = i\text{sgn}(\xi)e^{-a|\xi|}. \qquad (3.74)$$

Figure 3.3 shows an exponential function and its transform. Further transforms related to the exponential function can be generated by differentiating or integrating with respect to the parameter a. Through differentiation, we get

$$-\frac{dF_c}{da} = \mathcal{F}_c[xe^{-ax}] = \sqrt{\frac{2}{\pi}}\frac{a^2-\xi^2}{(a^2+\xi^2)^2}, \qquad (3.75)$$

$$-\frac{dF_s}{da} = \mathcal{F}_s[xe^{-ax}] = \sqrt{\frac{2}{\pi}}\frac{2a\xi}{(a^2+\xi^2)^2}, \qquad (3.76)$$

$$\mathcal{F}[|x|e^{-a|x|}] = \sqrt{\frac{2}{\pi}}\frac{a^2-\xi^2}{(a^2+\xi^2)^2}, \qquad (3.77)$$

$$\mathcal{F}[xe^{-a|x|}] = i\sqrt{\frac{2}{\pi}}\frac{2a\xi}{(a^2+\xi^2)^2}. \qquad (3.78)$$

Through integration, we get

$$\int_0^a F_c \, da = \mathcal{F}_c \left[\frac{1 - e^{-ax}}{x} \right] = \frac{1}{\sqrt{2\pi}} \log \frac{a^2 + \xi^2}{\xi^2}. \tag{3.79}$$

3.7.2 Gaussian Function

Consider the normal probability distribution function, which is also known as the Gaussian function, $\exp(-ax^2)$.

$$
\begin{aligned}
\mathcal{F}[e^{-ax^2}] &= \frac{1}{\sqrt{2\pi}} \int_{-\infty}^{\infty} e^{-ax^2 + i\xi x} \, dx \\
&= \frac{1}{\sqrt{2\pi}} \int_{-\infty}^{\infty} e^{-a[x^2 + 2i\xi x/(2a) + (i\xi/2a)^2] - \xi^2/4a} \, dx \\
&= \frac{e^{-\xi^2/4a}}{\sqrt{2\pi}} \int_{-\infty}^{\infty} e^{-a[x + i\xi x/(2a)]^2} \, dx \\
&= \frac{e^{-\xi^2/4a}}{\sqrt{2\pi}} \int_{-\infty}^{\infty} e^{-y^2} \frac{dy}{\sqrt{a}} \\
&= \frac{1}{\sqrt{2a}} e^{-\xi^2/4a},
\end{aligned}
\tag{3.80}
$$

where we have substituted

$$a(x + i\xi/2a)^2 = y^2, \qquad \int_{-\infty}^{\infty} e^{-y^2} \, dy = \sqrt{\pi}. \tag{3.81}$$

Also, if we look closely, we will see the integration path in the complex plane has been moved without crossing any singularities.

When $a = 1/2$,

$$\mathcal{F}[e^{-x^2/2}] = e^{-\xi^2/2}. \tag{3.82}$$

Functions with this symmetry are called self-reciprocal under the Fourier transform.

As this is an even function,

$$\mathcal{F}_c[e^{-ax^2}] = \frac{1}{\sqrt{2a}} e^{-\xi^2/4a}. \tag{3.83}$$

Differentiating this with ξ, we find

$$\mathcal{F}_s[xe^{-ax^2}] = \frac{1}{(2a)^{3/2}}\xi e^{-\xi^2/4a}, \tag{3.84}$$

which is self-reciprocal when $a = 1/2$.

Next, let us consider

$$\mathcal{F}[e^{iax^2}] = \frac{1}{\sqrt{2\pi}}\int_{-\infty}^{\infty} e^{i(ax^2+\xi x)}\, dx$$

$$= \frac{1}{\sqrt{2\pi}}\int_{-\infty}^{\infty} e^{ia(x+\xi/2a)^2 - i\xi^2/4a}\, dx$$

$$= \frac{1}{\sqrt{2\pi}}e^{-i\xi^2/4a}\int_{-\infty}^{\infty} e^{iax^2}\, dx, \tag{3.85}$$

where, in the last line, we replaced the variable $x + \xi/2a$ by x.

The integral

$$I = \int_{-\infty}^{\infty} e^{iax^2}\, dx = \frac{1}{\sqrt{a}}\int_{-\infty}^{\infty} e^{ix^2}\, dx. \tag{3.86}$$

In the complex plane, this is a line integral of e^{iz^2} along $z = x$. We may rotate this line by 45° to have $z = x + iy = (1+i)x$. Using this,

$$I = \frac{1+i}{\sqrt{a}}\int_{-\infty}^{\infty} e^{-2x^2}\, dx = \sqrt{\frac{\pi}{2a}}(1+i). \tag{3.87}$$

Then, Eq. (3.85) gives

$$\mathcal{F}[e^{iax^2}] = \frac{1+i}{2\sqrt{a}}e^{-i\xi^2/4a}.$$

For this even function, separating real and imaginary parts, we find

$$\mathcal{F}_c[\cos(ax^2)] = \frac{1}{2\sqrt{a}}\left[\cos\frac{\xi^2}{4a} + \sin\frac{\xi^2}{4a}\right], \tag{3.88}$$

$$\mathcal{F}_c[\sin(ax^2)] = \frac{1}{2\sqrt{a}}\left[\cos\frac{\xi^2}{4a} - \sin\frac{\xi^2}{4a}\right]. \tag{3.89}$$

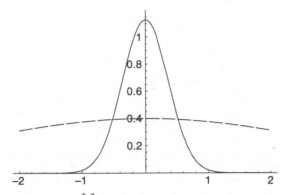

Figure 3.4. The function $ne^{-n^2x^2}/\sqrt{\pi}$ for $n=2$ and its Fourier transform (dashed curve).

In the Gaussian family, we may consider the delta sequence, $\delta_n = ne^{-n^2x^2}/\sqrt{\pi}$, which has the transform

$$\Delta_n(\xi) = \mathcal{F}[\delta_n(x)] = \frac{1}{\sqrt{2\pi}}e^{-\xi^2/4n^2}. \qquad (3.90)$$

Figure 3.4 shows a member of the delta sequence and its Fourier transform.

As $n \to \infty$, we get

$$\mathcal{F}[\delta(x)] = \frac{1}{\sqrt{2\pi}}. \qquad (3.91)$$

This has to be viewed as an extension of the Fourier transform beyond its application to absolutely integrable functions.

To relate the Fourier transform to the frequency content of a function, let us obtain the transform,

$$\mathcal{F}[e^{-a|x|}\cos\omega x] = \sqrt{\frac{2}{\pi}}\int_0^\infty e^{-ax}\cos\omega x\cos\xi x\,dx$$

$$= \frac{1}{\sqrt{2\pi}}\int_0^\infty e^{-ax}[\cos(\omega-\xi)+\cos(\omega+\xi)]\,dx$$

$$= \frac{1}{\sqrt{2\pi}}\left[\frac{a}{a^2+(\omega-\xi)^2}+\frac{a}{a^2+(\omega+\xi)^2}\right]. \qquad (3.92)$$

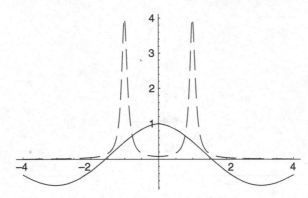

Figure 3.5. The function $e^{-a|x|}\cos\omega x$ and its Fourier transform (dashed curve) for $a = 0.1$ and $\omega = 1$.

A plot of the function and its transform are shown in Fig. 3.5. We see two spikes corresponding to $\xi = \omega$ and $\xi = -\omega$ in the transform. These spikes tend to delta functions as a approaches zero. We may verify one feature of the delta sequence, namely, the area under the curve is unity by examining the following relation concerning the inverse Fourier transform,

$$\frac{1}{\sqrt{2\pi}} \int_{-\infty}^{\infty} \sqrt{\frac{2}{\pi}} \frac{a}{a^2 + (\omega - \xi)^2} e^{-i(\xi - \omega)x} \, d\xi = e^{-a|x|}. \tag{3.93}$$

If we let $x = 0$, we get

$$\frac{1}{\pi} \int_{-\infty}^{\infty} \frac{a}{a^2 + (\omega - \xi)^2} \, d\xi = 1. \tag{3.94}$$

Defining

$$\delta_a(\xi - \omega) = \frac{1}{\pi} \frac{a}{a^2 + (\omega - \xi)^2}, \tag{3.95}$$

$$\mathcal{F}[e^{-a|x|}\cos\omega x] = \sqrt{\frac{\pi}{2}} [\delta_a(\xi - \omega) + \delta_a(\xi + \omega)], \tag{3.96}$$

$$\mathcal{F}[\cos\omega x] = \sqrt{\frac{\pi}{2}} [\delta(\xi - \omega) + \delta(\xi + \omega)]. \tag{3.97}$$

If a function is made up of a linear combination of $\cos \omega_n x$, $(n = 1, 2, \ldots, N)$ its Fourier transform will show delta functions located at $\pm \omega_n$. These are referred to as spectral lines.

3.7.3 Powers

In order to compute the Fourier transforms of functions of the form x^{-p}, first we introduce the Gamma function

$$\Gamma(x+1) = \int_0^\infty e^{-t} t^x \, dt. \tag{3.98}$$

Integrating by parts, we obtain the recurrence relation

$$\Gamma(x+1) = x\Gamma(x), \tag{3.99}$$

and, when x is an integer, n,

$$\Gamma(n+1) = n!. \tag{3.100}$$

Let us evaluate

$$F_c + iF_s = \sqrt{\frac{2}{\pi}} \int_0^\infty x^{-p} e^{i\xi x} \, dx. \tag{3.101}$$

For x^{-p} to go to zero at infinity, $p > 0$. In the complex plane, $z = x + iy$, this is a line integral along $C : z = x$. We may distort the contour as shown in Fig. 3.6 without including any singular points. Then

$$\int_C = \int_{C_0} + \int_{C_1} + \int_{C_\infty}, \tag{3.102}$$

with C_0 having a small radius ϵ and C_∞ a large radius R. First, let us consider the integral over C_1:

$$\begin{aligned}
\int_{C_1} &= \sqrt{\frac{2}{\pi}} \int_\epsilon^R e^{-i\pi p/2} y^{-p} e^{-\xi y} i \, dy \\
&= \sqrt{\frac{2}{\pi}} \xi^{p-1} e^{i\pi(1-p)/2} \int_0^\infty y^{-p} e^{-y} \, dy \\
&= \sqrt{\frac{2}{\pi}} \xi^{p-1} e^{i\pi(1-p)/2} \Gamma(1-p), \tag{3.103}
\end{aligned}$$

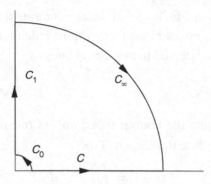

Figure 3.6. Two paths for evaluating the integral of $z^{-p}\exp(i\xi z)$.

where we have taken the limits $\epsilon \to 0$ and $R \to \infty$. The integral over C_0

$$\int_{C_0} \le \sqrt{\frac{2}{\pi}}\left|\int_0^{\pi/2} \epsilon^{1-p}e^{-i\theta p-\xi\epsilon\sin\theta}\,d\theta\right|. \tag{3.104}$$

As $\epsilon \to 0$, this integral vanishes if $p < 1$. Thus, $0 < p < 1$. The integral over C_∞

$$\int_{C_\infty} \le \sqrt{\frac{2}{\pi}}\left|\int_{\pi/2}^0 R^{1-p}e^{-i\theta p-\xi R\sin\theta}\,d\theta\right|. \tag{3.105}$$

As $\xi > 0$, in the limit $R \to \infty$, this integral goes to zero. Thus,

$$F_c + iF_s = \sqrt{\frac{2}{\pi}}\xi^{p-1}\Gamma(1-p)e^{i(1-p)\pi/2}. \tag{3.106}$$

For an inverse transform to exist, this expression has to be absolutely integrable, which also confirms $1 - p < 1$ or $0 < p < 1$.

Separating the real and the imaginary parts and using the notation $q = 1 - p$, we get

$$F_c = \sqrt{\frac{2}{\pi}}\xi^{-q}\Gamma(q)\cos\pi q/2,$$

$$F_s = \sqrt{\frac{2}{\pi}}\xi^{-q}\Gamma(q)\sin\pi q/2. \tag{3.107}$$

When $p = 1/2 = q$, using $\Gamma(1/2) = \sqrt{\pi}$, we obtain the self-reciprocal relations

$$\mathcal{F}_c\left[\frac{1}{\sqrt{x}}\right] = \frac{1}{\sqrt{\xi}}, \quad \mathcal{F}_s\left[\frac{1}{\sqrt{x}}\right] = \frac{1}{\sqrt{\xi}}. \tag{3.108}$$

Using the even and odd extensions,

$$\mathcal{F}[|x|^{-p}] = \sqrt{\frac{2}{\pi}}|\xi|^{-q}\Gamma(q)\cos\pi q/2, \tag{3.109}$$

$$\mathcal{F}[\mathrm{sgn}(x)|x|^{-p}] = \sqrt{\frac{2}{\pi}}\mathrm{sgn}(\xi)|\xi|^{-q}\Gamma(q)\sin\pi q/2. \tag{3.110}$$

3.8 CONVOLUTION INTEGRAL

Under the Fourier transform, the convolution integral of two functions, f and g, is defined as

$$f * g(x) = \frac{1}{\sqrt{2\pi}}\int_{-\infty}^{\infty} f(x-t)g(t)\,dt, \tag{3.111}$$

where the operator $*$ is used to denote convolution. Using

$$x - t = \tau, \quad t = x - \tau, \tag{3.112}$$

we see

$$f * g(x) = \frac{1}{\sqrt{2\pi}}\int_{-\infty}^{\infty} f(\tau)g(x-\tau)\,dt = g * f(x), \tag{3.113}$$

which shows the commutative property of the convolution operator. The Fourier transform of the convolution integral,

$$\begin{aligned}
\mathcal{F}[f * g] &= \frac{1}{2\pi}\int_{-\infty}^{\infty}\int_{-\infty}^{\infty} f(x-t)g(t)e^{i\xi x}\,dt\,dx \\
&= \frac{1}{2\pi}\int_{-\infty}^{\infty}\int_{-\infty}^{\infty} f(\tau)g(t)e^{i\xi(t+\tau)}\,dt d\tau, \quad x-t=\tau, \\
&= F(\xi)G(\xi). \tag{3.114}
\end{aligned}$$

This relation shows that if a Fourier transform is factored as F times G, we may invert them individually, to get f and g and then convolute them to obtain the inverse of FG. It also implies that the convolution is defined to suit the transform.

3.8.1 Inner Products and Norms

An important property of the Fourier transform is that it conserves the inner products and norms. From the inversion formula,

$$f * h(x) = \mathcal{F}^{-1}[F(\xi)H(\xi)], \tag{3.115}$$

where we assume f and h are real, letting $x = 0$,

$$\int_{-\infty}^{\infty} f(t)h(-t)\, dt = \int_{-\infty}^{\infty} F(\xi)H(\xi)\, d\xi. \tag{3.116}$$

If $g^*(t) = h(-t)$,

$$H(\xi) = \frac{1}{\sqrt{2\pi}} \int_{-\infty}^{\infty} g^*(-t)e^{i\xi t}\, dt,$$

$$= \frac{1}{\sqrt{2\pi}} \int_{-\infty}^{\infty} g^*(t)e^{-i\xi t}\, dt = G^*(\xi). \tag{3.117}$$

Then

$$\langle f, g \rangle = \langle F, G \rangle, \tag{3.118}$$

where the inner products are defined as

$$\langle f, g \rangle \equiv \int_{-\infty}^{\infty} fg^*\, dx, \quad \langle F, G \rangle \equiv \int_{-\infty}^{\infty} F(\xi)G^*(\xi)\, d\xi. \tag{3.119}$$

The choice, $g = f$, leads to the conservation relation for the norm:

$$\|f\| = \|F\|. \tag{3.120}$$

In physical applications, often, the norm is related to the energy content and Eq. (3.120), known as the Parseval's relation, shows the same energy content in the frequency domain. A plot of $\|F\|^2$ as a function of ξ is called a power spectrum of a signal $f(x)$.

3.8.2 Convolution for Trigonometric Transforms

The convolution integral for trigonometric transforms are not as simple as for the Fourier transforms. We may obtain them by considering

$$\int_0^\infty F_c G_c \cos\xi x\, d\xi = \sqrt{\frac{2}{\pi}}\int_0^\infty\int_0^\infty G_c f(t)\cos\xi t\cos\xi x\, dt\, d\xi$$

$$= \frac{1}{\sqrt{2\pi}}\int_0^\infty\int_0^\infty f(t)G_c[\cos\xi(t-x)$$

$$+\cos\xi(t+x)]\,d\xi\, dt$$

$$= \frac{1}{2}\int_0^\infty f(t)[g(|t-x|)+g(t+x)]\,dt. \qquad (3.121)$$

Similarly,

$$\int_0^\infty F_s G_s \cos\xi x\, d\xi = \sqrt{\frac{2}{\pi}}\int_0^\infty\int_0^\infty G_s f(t)\sin\xi t\cos\xi x\, dt\, d\xi$$

$$= \frac{1}{\sqrt{2\pi}}\int_0^\infty\int_0^\infty f(t)G_s[\sin\xi(t+x)$$

$$+\sin\xi(t-x)]\,d\xi\, dt$$

$$= \frac{1}{2}\int_0^\infty f(t)[\operatorname{sgn}(t-x)g(|t-x|)+g(x+t)]\,dt,$$

$$(3.122)$$

$$\int_0^\infty F_s G_c \sin\xi x\, d\xi = \sqrt{\frac{2}{\pi}}\int_0^\infty\int_0^\infty G_c f(t)\sin\xi t\sin\xi x\, dt\, d\xi$$

$$= \frac{1}{\sqrt{2\pi}}\int_0^\infty\int_0^\infty f(t)G_c[\cos\xi(t-x)$$

$$-\cos\xi(t+x)]\,d\xi\, dt$$

$$= \frac{1}{2}\int_0^\infty f(t)[g(|t-x|)-g(t+x)]\,dt. \qquad (3.123)$$

Setting $x=0$ in Eqs. (3.121), and (3.122), we obtain the inner product conservation relations

$$\langle F_c, G_c\rangle = \langle f,g\rangle, \quad \langle F_s, G_s\rangle = \langle f,g\rangle, \qquad (3.124)$$

where all functions involved are real and the inner products are defined using integrals from 0 to ∞.

The special case, $g = f$, gives

$$\|F_c\| = \|f\|, \quad \|F_s\| = \|f\|. \tag{3.125}$$

3.9 MIXED TRIGONOMETRIC TRANSFORM

As we have seen, if $f'(0)$ is given, we use the Fourier Cosine transform, and if $f(0)$ is given, we use the Fourier Sine transform. There are cases where a linear combination of $f'(0)$ and $f(0)$ is given. As an example, let us consider the Newton's law of cooling at the surface, $x = 0$, of a body extending to infinity,

$$f'(0) - hf(0) = 0, \tag{3.126}$$

where $f(x)$ is the relative temperature, $T(x) - T_\infty$ and h is a constant. This boundary condition is known as the radiation condition in the literature.

In order to obtain the preceding combination of boundary terms, let us consider the mixed trigonometric transform,

$$\mathcal{F}_R[f] = F_R(\xi) = \sqrt{\frac{2}{\pi}} \int_0^\infty f(x)[a\cos x\xi + b\sin x\xi]\,dx, \tag{3.127}$$

where a and b have to be chosen to obtain the boundary term as in the radiation condition of Eq. (3.126) from the transform of $f''(x)$. The subscript, R, is used to indicate the radiation condition. We have

$$\mathcal{F}_R[f''] = -\sqrt{\frac{2}{\pi}}\left[af'(0) - b\xi f(0)\right] - \xi^2 F_R(\xi). \tag{3.128}$$

Choosing

$$a = \xi, \quad b = h, \tag{3.129}$$

we recover the left-hand side of Eq. (3.126). Our definition of the mixed transform is

$$\mathcal{F}_R[f] = F_R(\xi) = \sqrt{\frac{2}{\pi}} \int_0^{\infty} f(x)[\xi \cos x\xi + h \sin x\xi]\,dx. \qquad (3.130)$$

The mixed transform can also be written as the Sine or the Cosine transform of a linear combination f and its derivative or anti-derivative. Using the Sine transform, we obtain an expression for the inverse of the mixed transform. Consider

$$F_R(\xi) = \sqrt{\frac{2}{\pi}} \int_0^{\infty} f(x)\left[\frac{d}{dx} + h\right]\sin x\xi\,dx$$

$$= \sqrt{\frac{2}{\pi}} f(x)\sin x\xi \Big|_0^{\infty} - \sqrt{\frac{2}{\pi}} \int_0^{\infty} [f'(x) - hf(x)]\sin x\xi\,dx$$

$$= -\mathcal{F}_s[f' - hf],$$

$$f'(x) - hf(x) = -\mathcal{F}_s^{-1}[F_R]. \qquad (3.131)$$

This is a first-order differential equation, which can be solved to get $f(x)$. The convolution theorem for the mixed transform states that

$$\mathcal{F}_R[f * g] = \xi^{-1} F_R G_R, \qquad (3.132)$$

where the convolution integral $f * g$ is defined as

$$f * g(x) = \frac{1}{\sqrt{2\pi}} \int_0^{\infty} g(t)\left[f(x+t) + f(|x-t|) + h \int_{|x-t|}^{x+t} f(\tau)\,d\tau\right]dt. \qquad (3.133)$$

For a derivation of this result, the reader may consult Sneddon (1972). For a solution of problems involving the radiation boundary condition, we may use Eq. (3.131) to invert the transform.

3.9.1 Example: Mixed Transform

To find the mixed transform

$$\mathcal{F}_R[e^{-ax}], \qquad (3.134)$$

we use

$$F_c = \sqrt{\frac{2}{\pi}} \frac{a}{a^2 + \xi^2}, \quad F_s = \sqrt{\frac{2}{\pi}} \frac{\xi}{a^2 + \xi^2}, \tag{3.135}$$

to get

$$F_R = \xi F_c + h F_s = \sqrt{\frac{2}{\pi}} \frac{a+h}{a^2 + \xi^2} \xi. \tag{3.136}$$

3.10 MULTIPLE FOURIER TRANSFORMS

For functions of two variables, x and y, if both variables extend from $-\infty$ to ∞, we may take the Fourier transform with respect to x and y sequentially, to have

$$F(\xi, \eta) = \frac{1}{2\pi} \int_{-\infty}^{\infty} \int_{-\infty}^{\infty} f(x, y) e^{i(x\xi + y\eta)} \, dx \, dy. \tag{3.137}$$

In an n–dimensional space Ω, if

$$x = x_1, x_2, \ldots, x_n, \quad \boldsymbol{\xi} = \xi_1, \xi_2, \ldots, \xi_n, \tag{3.138}$$

we have

$$x_1 \xi_1 + x_2 \xi_2 + \cdots + x_n \xi_n = \boldsymbol{x} . \boldsymbol{\xi}. \tag{3.139}$$

With this, the generalized n–dimensional transform can be written as

$$F(\boldsymbol{\xi}) = \frac{1}{(2\pi)^{n/2}} \int_{\Omega} f(\boldsymbol{x}) e^{i\boldsymbol{x} . \boldsymbol{\xi}} \, d\Omega. \tag{3.140}$$

In engineering applications, n has the value of two or three.

3.11 APPLICATIONS OF FOURIER TRANSFORM

In the following subsections we discuss some examples of solutions of differential and integral equations obtained using the Fourier transform.

3.11.1 Examples: Partial Differential Equations

1. Laplace Equation in the Half Space

Consider an incompressible, irrotational, two-dimensional flow of a fluid in the upper half plane: $-\infty < x < \infty$, $0 < y < \infty$ (see Fig. 3.7).

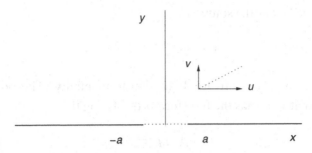

Figure 3.7. Semi-infinite half space with prescribed vertical fluid flow in $-a < x < a$.

Let u and v denote the components of velocity along the x- and the y-axis, respectively. We assume the fluid is flowing into this domain with $v = f(x)$, $u = 0$ along the slot, $-a < x < a$, $y = 0$. Both components of velocity vanish at infinity.

For incompressibility

$$\frac{\partial u}{\partial x} + \frac{\partial v}{\partial y} = 0, \tag{3.141}$$

and for irrotationality

$$\frac{\partial u}{\partial y} - \frac{\partial v}{\partial x} = 0. \tag{3.142}$$

Differentiating Eq. (3.141) with respect to x and Eq. (3.142) with respect to y and adding the two, we get

$$\nabla^2 u = 0. \tag{3.143}$$

Similarly, eliminating u, we have

$$\nabla^2 v = 0. \tag{3.144}$$

Thus, the velocity components satisfy the Laplace equation. Taking the Fourier transform of Eq. (3.144) with respect to x,

$$V(\xi, y) = \mathcal{F}[v(x,y), x \to \xi], \tag{3.145}$$

we have

$$\frac{d^2 V}{dy^2} - \xi^2 V = 0. \tag{3.146}$$

This equation has the solution

$$V = Ae^{-|\xi|y} + Be^{|\xi|y}, \tag{3.147}$$

where B has to be zero for V to vanish at infinity. The boundary condition at $y = 0$ has the transform $F(\xi)$. Using this

$$A = F(\xi), \tag{3.148}$$

and

$$V(\xi, y) = F(\xi)e^{-|\xi|y}. \tag{3.149}$$

Using the convolution integral and

$$\mathcal{F}^{-1}[e^{-|\xi|y}] = \sqrt{\frac{2}{\pi}} \frac{y}{x^2 + y^2}, \tag{3.150}$$

we invert this to get

$$v(x, y) = \frac{y}{\pi} \int_{-\infty}^{\infty} \frac{f(t)dt}{(x - t)^2 + y^2}. \tag{3.151}$$

To find u, we take the transform of Eq. (3.141),

$$-i\xi U = |\xi| F(\xi)e^{-|\xi|y}, \tag{3.152}$$

or

$$U = i \operatorname{sgn} \xi F(\xi)e^{-|\xi|y}. \tag{3.153}$$

Using

$$\mathcal{F}^{-1}[i \operatorname{sgn} \xi e^{-|\xi|y}] = \sqrt{\frac{2}{\pi}} \frac{x}{x^2 + y^2}, \tag{3.154}$$

we find

$$u(x, y) = \frac{1}{\pi} \int_{-\infty}^{\infty} \frac{(x - t)f(t)dt}{(x - t)^2 + y^2}. \tag{3.155}$$

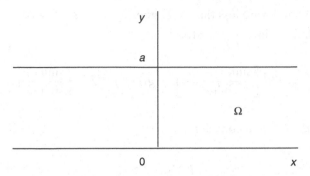

Figure 3.8. Infinite strip with prescribed temperatures at $y = 0$ and $y = a$.

2. Steady-State Temperature Distribution in a Strip

We consider a strip Ω: $-\infty < x < \infty$, $\quad 0 < y < a$, (see Fig. 3.8). The relative temperature $u(x,y)$ satisfies

$$\nabla^2 u = 0 \quad \text{in} \quad \Omega, \tag{3.156}$$

$$u(x,0) = f_0(x), \quad u(x,a) = f_1(x). \tag{3.157}$$

Taking the Fourier transform of the Laplace equation,

$$\frac{d^2 U}{dy^2} - \xi^2 U = 0, \tag{3.158}$$

where

$$U(\xi,y) = \mathcal{F}[u(x,y), x \to \xi]. \tag{3.159}$$

Unlike in the case of the semi-infinite domain, for the strip it is convenient to write the solution in the form

$$U = A \sinh(a - y)\xi + B \sinh y\xi. \tag{3.160}$$

With the Fourier transforms of the boundary conditions, we obtain

$$A = F_0(\xi)/\sinh a\xi, \quad B = F_1(\xi)/\sinh a\xi. \tag{3.161}$$

Then

$$U = F_0(\xi)\frac{\sinh(a - y)\xi}{\sinh a\xi} + F_1(\xi)\frac{\sinh y\xi}{\sinh a\xi}. \tag{3.162}$$

As $(a-y)$ and y are less than a, U goes to zero as $|\xi| \to \infty$.

To find the inverse transforms,

$$g_0(x,y) = \mathcal{F}^{-1}\left[\frac{\sinh(a-y)\xi}{\sinh a\xi}\right], \quad g_1(x,y) = \mathcal{F}^{-1}\left[\frac{\sinh y\xi}{\sinh a\xi}\right], \quad (3.163)$$

we use $a\xi \to \xi$, and consider

$$I(x,c) = \int_{-\infty}^{\infty} \frac{e^{c\xi - ix\xi}}{\sinh \xi}\, d\xi, \quad c < 1. \quad (3.164)$$

In terms of $I(x,c)$,

$$g_0(x,y) = \frac{1}{\sqrt{8\pi a}}[I(x/a,\bar{y}/a) - I(x/a,-\bar{y}/a)], \quad (3.165)$$

$$g_1(x,y) = \frac{1}{\sqrt{8\pi a}}[I(x/a,y/a) - I(x/a,-y/a)], \quad (3.166)$$

where $\bar{y} = a - y$. Using the convolution theorem, the solution can be written as

$$u(x,y) = \frac{1}{\sqrt{2\pi}}\int_{-\infty}^{\infty} [g_0(x-t,y)f_0(t) + g_1(x-t,y)f_1(t)]\, dt. \quad (3.167)$$

Here, g_0 and g_1 are the Green's functions associated with the nonhomogeneous boundary conditions.

To evaluate the integral I, we use the contour shown in Fig. 3.9. In the complex $\zeta = \xi + i\eta$ plane, the function, $f(\zeta) = e^{(c-ix)\zeta}/\sinh \zeta$, has simple poles at $\zeta = 0, \pi i, 2\pi i$. The integrals over C and C_1 add up to $2\pi i$ times the sum of half of the residues at 0 and $2\pi i$ and a full residue at πi with the integrals at infinity vanishing, in which case.

$$\int_C f(\zeta)d\zeta = \int_{-\infty}^{\infty} f(\xi)d\xi = I. \quad (3.168)$$

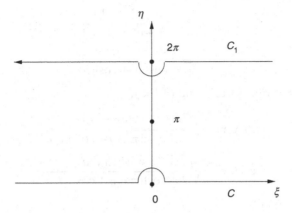

Figure 3.9. Contour in the complex ζ plane.

On C_1, $\zeta = \xi + 2\pi i$ and

$$\int_{C_1} f(\zeta)d\zeta = \int_{\infty}^{-\infty} \frac{e^{2\pi(ic+x)+(c-ix)\xi}}{\sinh(\xi + 2\pi i)} d\xi$$

$$= -e^{2\pi(ic+x)} \int_{-\infty}^{\infty} f(\xi)d\xi$$

$$= -e^{2\pi(ic+x)} I. \qquad (3.169)$$

The residues are

$$\text{Res}(0) = 1,$$

$$\text{Res}(\pi i) = -e^{\pi(ic+x)},$$

$$\text{Res}(2\pi i) = e^{2\pi(ic+x)}.$$

As the points $\zeta = 0$ and $\zeta = 2\pi i$ are on the contour, the Cauchy principal values of the integrals through them give half the residues. Thus,

$$(1 - e^{2\pi(ic+x)})I = \pi i(1 + e^{2\pi(ic+x)} - 2e^{\pi(ic+x)}),$$

$$I = \pi i \frac{(1 - e^{\pi(ic+x)})^2}{(1 - e^{\pi(ic+x)})(1 + e^{\pi(ic+x)})}$$

$$= \pi i \frac{1 - e^{\pi(ic+x)}}{1 + e^{\pi(ic+x)}} \frac{1 + e^{\pi(-ic+x)}}{1 + e^{\pi(-ic+x)}}$$

$$= \pi \frac{\sin \pi c - i \sinh \pi x}{\cos \pi c + \cosh \pi c}. \tag{3.170}$$

Using this in Eqs. (3.165) and (3.166),

$$g_0(x,y) = \sqrt{\frac{\pi}{2a^2}} \frac{\sin \pi \bar{y}/a}{\cosh \pi x/a + \cos \pi \bar{y}/a}, \tag{3.171}$$

$$g_1(x,y) = \sqrt{\frac{\pi}{2a^2}} \frac{\sin \pi y/a}{\cosh \pi x/a + \cos \pi y/a}, \tag{3.172}$$

where $\bar{y} = a - y$.

3. Transient Heat Conduction in a Semi-infinite Rod

Consider a semi-infinite rod occupying $0 < x < \infty$. The transient relative temperature $u(x,t)$ satisfies

$$\kappa \frac{\partial^2 u}{\partial x^2} = \frac{\partial u}{\partial t}, \tag{3.173}$$

where κ is the diffusivity. The boundary conditions are

$$u(0,t) = 0, \quad u \to 0 \quad \text{as} \quad x \to \infty. \tag{3.174}$$

The initial temperature distribution is given as

$$u(x,0) = f(x), \tag{3.175}$$

with $f(0) = 0$ and $f \to 0$ as $x \to \infty$. For a semi-infinite domain with $u(0,t)$ prescribed, we use the Fourier Sine transform.

$$U(\xi,t) = \mathcal{F}_s[u(x,t)], \quad F_s(\xi) = \mathcal{F}_s[f(x)], \tag{3.176}$$

$$\mathcal{F}_s\left[\frac{\partial^2 u}{\partial x^2}\right] = -\xi^2 U. \tag{3.177}$$

Taking the transform of Eq. (3.173), we have

$$\frac{dU}{dt} + \kappa \xi^2 U = 0, \tag{3.178}$$

which has the solution

$$U(\xi,t) = Ae^{-\kappa t\xi^2},\tag{3.179}$$

where A is found from the initial condition, $U(\xi,0) = A = F_s(\xi)$. Thus,

$$U(\xi,t) = F_s(\xi)e^{-\kappa t\xi^2}.\tag{3.180}$$

For special initial temperature distributions, such as

$$f(x) = u_0 x e^{-x^2/(4a^2)},\tag{3.181}$$

we have from Eq. (3.84)

$$F(\xi) = u_0(2\sqrt{a})^3\xi e^{-a^2\xi^2},\tag{3.182}$$

$$U(\xi,t) = u_0(2\sqrt{a})^3\xi e^{-(a^2+\kappa t)\xi^2}.\tag{3.183}$$

Inverting this gives

$$u(x,t) = \frac{u_0 x}{(1+\kappa t/a^2)^{3/2}}e^{-x^2/[4(a^2+\kappa t)]}.\tag{3.184}$$

Figure 3.10 shows the peak temperature diminishing and moving to the right as time passes.

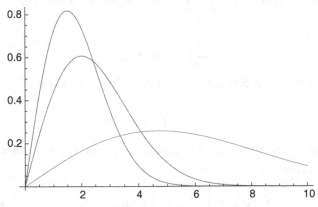

Figure 3.10. Temperature distributions u/u_0 for $t = 0.1, 1.0, 10.0$ when $a^2 = \kappa = 1$.

In a more general scenario, we would like to keep $f(x)$ arbitrary and to obtain the Green's function to write the solution in a convolutional form. To this end, from Eq. (3.123), we have

$$\mathcal{F}_s^{-1}[F_s G_c] = \frac{1}{\sqrt{2\pi}} \int_0^\infty f(t)[g(|x-t|) - g(x+t)]\,dt, \qquad (3.185)$$

where

$$G_c = e^{-\kappa t \xi^2}, \quad g(x) = \frac{1}{\sqrt{2\kappa t}} e^{-x^2/(4\kappa t)}. \qquad (3.186)$$

Thus,

$$u(x,t) = \frac{1}{\sqrt{4\pi\kappa t}} \int_0^\infty f(x')[e^{-(x-x')^2/(4\kappa t)} - e^{-(x+x')^2/(4\kappa t)}]\,dx'. \qquad (3.187)$$

4. Transient Heat Conduction in a Rod with Radiation Condition

Here we reconsider the previous heat conduction problem with the boundary condition at $x = 0$ changed to the radiation condition (Newton's law of cooling):

$$\frac{\partial u}{\partial x}(0,t) - hu(0,t) = 0. \qquad (3.188)$$

Using the mixed trigonometric transform,

$$U_R(\xi,t) = \mathcal{F}_R[u(x,t)] = \sqrt{\frac{2}{\pi}} \int_0^\infty u(x,t)[\xi\cos\xi x + h\sin\xi x]\,dx, \qquad (3.189)$$

we obtain

$$\frac{\partial U_R}{\partial t} = -\kappa\xi^2 U_R. \qquad (3.190)$$

The solution of this equation satisfying the initial condition, $U_R(\xi,0) = F_R(\xi)$, is

$$U_R(\xi,t) = F_R(\xi)e^{-\kappa t \xi^2}. \qquad (3.191)$$

Using the relation (3.131) between the mixed transform and the Sine transform,

$$U_R = -\mathcal{F}_s\left[\frac{\partial u}{\partial x} - hu\right], \quad F_R = -\mathcal{F}_s[g], \qquad (3.192)$$

where
$$g \equiv \frac{\partial f}{\partial x} - hf. \tag{3.193}$$

Then
$$\frac{\partial u}{\partial x} - hu = \mathcal{F}_s^{-1} \left[\sqrt{\frac{2}{\pi}} e^{-\kappa t \xi^2} \int_0^\infty g(x') \sin \xi x' \, dx' \right]$$

$$= \frac{2}{\pi} \int_0^\infty \int_0^\infty g(x') e^{-\kappa t \xi^2} \sin \xi x \sin \xi x' \, dx' \, d\xi$$

$$= \frac{1}{\pi} \int_0^\infty \int_0^\infty g(x') e^{-\kappa t \xi^2} [\cos(x - x')\xi - \cos(x + x')\xi] \, d\xi \, dx'$$

$$= \frac{1}{\sqrt{4\pi\kappa t}} \int_0^\infty g(x') [e^{-(x-x')^2/(4\kappa t)} - e^{-(x+x')^2/(4\kappa t)}] \, dx',$$

$$u = \frac{1}{\sqrt{4\pi\kappa t}} \int_0^\infty g(x') \left(\frac{\partial}{\partial x} - h \right)^{-1} [e^{-(x-x')^2/(4\kappa t)} - e^{-(x+x')^2/(4\kappa t)}] \, dx'. \tag{3.194}$$

Next, we want to obtain the result of the inverse differential operator acting on the x-dependent Gaussian functions inside the integral. This is equivalent to seeking a solution of the differential equation

$$\frac{\partial v}{\partial x} - hv = e^{-(x-x')^2/(4\kappa t)} - e^{-(x+x')^2/(4\kappa t)}. \tag{3.195}$$

Note that $v(x)$ satisfies the radiation boundary condition at $x = 0$. We may solve for v using superposition by considering only the first term on the right-hand side. Using an integrating factor,

$$\frac{\partial(e^{-hx}v)}{\partial x} = \exp \left[-\frac{x^2 - 2xa + 4\kappa htx + a^2}{4\kappa t} \right], \tag{3.196}$$

where $a = \pm x'$. Completing the square,

$$\frac{\partial(e^{-hx}v)}{\partial x} = \exp \left[\frac{(a - 2\kappa ht)^2 - a^2}{4\kappa t} \right] \exp \left[-\frac{(x - a + 2\kappa ht)^2}{4\kappa t} \right]$$

$$= \exp(-ha + \kappa h^2 t) \exp \left[-\frac{(x - a + 2\kappa ht)^2}{4\kappa t} \right]. \tag{3.197}$$

Integrating both sides, we obtain

$$v = \sqrt{\pi \kappa t}\exp[h(x-a+\kappa ht)]\mathrm{erf}\left[\frac{x-a+2\kappa ht}{\sqrt{4\kappa t}}\right]. \tag{3.198}$$

With this, superimposing the two expressions for v corresponding to $a = \pm x'$, the solution given in Eq. (3.194) can be written as

$$u = \frac{1}{2}\int_0^\infty g(x')[w(x-x')-w(x+x')]dx', \tag{3.199}$$

where

$$w(x) = e^{h(x+\kappa ht)}\mathrm{erf}\left[\frac{x+2\kappa ht}{\sqrt{4\kappa t}}\right]. \tag{3.200}$$

5. Transient Heat Conduction in an Infinite Plate

The diffusion equation and the initial condition for the relative temperature u are

$$\kappa \nabla^2 u(x,y,t) = \frac{\partial u}{\partial t}, \quad u(x,y,0) = f(x,y), \tag{3.201}$$

where f and u go to zero as $\sqrt{x^2+y^2} \to \infty$.

Using the double Fourier transform,

$$\mathcal{F}[u(x,y,t), x \to \xi, y \to \eta] = U(\xi,\eta,t), \tag{3.202}$$

$$\mathcal{F}[f(x,y), x \to \xi, y \to \eta] = F(\xi,\eta), \tag{3.203}$$

the differential equation can be written as

$$\frac{dU}{dt} = -\kappa(\xi^2+\eta^2)U, \tag{3.204}$$

which has the solution satisfying the given initial condition,

$$u = F(\xi,\eta)e^{-\kappa t(\xi^2+\eta^2)}. \tag{3.205}$$

The inverse transform of the exponential term can be found by considering it as a product of a function of ξ and a function of η.

$$g(x,y,t) = \mathcal{F}^{-1}[e^{-\kappa t(\xi^2+\eta^2)}] = \frac{1}{2\kappa t}e^{-(x^2+y^2)/(4\kappa t)}. \tag{3.206}$$

Using the convolution theorem, the solution can be written as

$$u(x,y,t) = \frac{1}{4\pi\kappa t} \int_{-\infty}^{\infty} \int_{-\infty}^{\infty} f(x',y') \exp\left[-\frac{(x-x')^2+(y-y')^2}{4\kappa t}\right] dx' dy'.$$
(3.207)

6. Laplace Equation in a Semi-infinite 3D Domain

Consider the equation

$$\frac{\partial^2 u}{\partial x^2} + \frac{\partial^2 u}{\partial y^2} + \frac{\partial^2 u}{\partial z^2} = 0,$$
(3.208)

with the conditions

$$u(x,y,0) = f(x,y), \quad u \to 0 \quad \text{as} \quad R \equiv \sqrt{x^2+y^2+z^2} \to \infty. \quad (3.209)$$

Taking the double Fourier transform with $x \to \xi$ and $y \to \eta$, we get

$$\frac{\partial^2 U}{\partial z^2} - (\xi^2 + \eta^2)U = 0,$$
(3.210)

where

$$U(\xi,\eta,z) = \mathcal{F}[u, x \to \xi, y \to \eta].$$
(3.211)

The boundary condition becomes

$$U(\xi,\eta,0) = F(\xi,\eta).$$
(3.212)

The solution of the differential equation (3.210) satisfying the boundary condition at $z = 0$ and the condition at infinity is

$$U(\xi,\eta,z) = F(\xi,\eta)e^{-\sqrt{\xi^2+\eta^2}z}.$$
(3.213)

We may invert this product by using convolution after obtaining the inverse of each factor. To this end, let

$$g(x,y,z) = \mathcal{F}^{-1}[e^{-\sqrt{\xi^2+\eta^2}z}]$$

$$= \frac{1}{2\pi} \int_{-\infty}^{\infty} \int_{-\infty}^{\infty} e^{-\sqrt{\xi^2+\eta^2}z} e^{-i(x\xi+y\eta)} d\xi d\eta. \quad (3.214)$$

We may transform this integral using the polar coordinates

$$\xi = \rho \cos \phi, \quad \eta = \rho \sin \phi, \tag{3.215}$$

$$x = r \cos \theta, \quad y = r \sin \theta. \tag{3.216}$$

Then

$$g = \frac{1}{2\pi} \int_0^{2\pi} \int_0^\infty e^{-[z + ir \cos(\phi - \theta)]\rho} \rho \, d\rho \, d\phi. \tag{3.217}$$

The integral with respect to ϕ from 0 to 2π can be changed to the limits, θ to $2\pi + \theta$. Using a new angle

$$\psi = \phi - \theta, \tag{3.218}$$

we have

$$
\begin{aligned}
g &= \frac{1}{2\pi} \int_0^{2\pi} \int_0^\infty e^{-[z + ir \cos \psi]\rho} \rho \, d\rho \, d\psi \\
&= \frac{1}{2\pi} \int_0^{2\pi} \left[-\frac{\rho e^{-[z + ir \cos \psi]\rho}}{z + ir \cos \psi} - \frac{e^{-[z + ir \cos \psi]\rho}}{(z + ir \cos \psi)^2} \right]_0^\infty d\psi \\
&= \frac{1}{2\pi} \int_0^{2\pi} \frac{d\psi}{(z + ir \cos \psi)^2} \\
&= \frac{dI}{dz},
\end{aligned}
\tag{3.219}
$$

where

$$I = -\frac{1}{2\pi} \int_0^{2\pi} \frac{d\psi}{z + ir \cos \psi}. \tag{3.220}$$

This integral is evaluated using

$$\zeta = e^{i\psi}, \quad d\zeta = i e^{i\psi} d\psi, \quad \cos \psi = (\zeta + \zeta^{-1})/2, \tag{3.221}$$

and the unit circle, $|\zeta| = 1$, as the closed contour. Thus,

$$
\begin{aligned}
I &= -\frac{1}{2\pi} \oint_{|\zeta|=1} \frac{1}{z + ir(\zeta + \zeta^{-1})/2} \frac{d\zeta}{i\zeta} \\
&= \frac{1}{\pi r} \oint_{|\zeta|=1} \frac{d\zeta}{\zeta^2 + 1 + 2z\zeta/(ir)}.
\end{aligned}
\tag{3.222}
$$

The integrand has simple poles at

$$\zeta = iz/r \pm i\sqrt{1 + z^2/r^2};$$ (3.223)

only one of the poles is inside the contour. Using the residue theorem,

$$I = -\frac{1}{r}\sqrt{1 + z^2/r^2},$$ (3.224)

and

$$g = \frac{z}{(z^2 + r^2)^{3/2}}.$$ (3.225)

The solution of the Laplace equation can be written as

$$u(x,y,z) = \frac{z}{2\pi} \int_{-\infty}^{\infty}\int_{-\infty}^{\infty} \frac{f(x',y')dx'dy'}{[(x-x')^2 + (y-y')^2 + z^2]^{3/2}}.$$ (3.226)

3.11.2 Examples: Integral Equations

1. The Fourier Integral

The Fourier transform

$$\frac{1}{\sqrt{2\pi}} \int_{-\infty}^{\infty} f(x)e^{ix\xi}\,dx = F(\xi)$$ (3.227)

can be considered as a singular integral equation with kernel, $k(x,\xi) = (2\pi)^{-1/2}\exp(ix\xi)$, with a given function $F(\xi)$ on the right-hand side. By the Fourier integral theorem, the solution of this equation is

$$f(x) = \frac{1}{\sqrt{2\pi}} \int_{-\infty}^{\infty} F(\xi)e^{-ix\xi}\,d\xi.$$ (3.228)

In this context, the self-reciprocal functions we have seen earlier become eigenfunctions with eigenvalues of unity for our integral operator. This illustrates a peculiar property of singular integral equations that a particular eigenvalue can have infinitely many eigenfunctions.

From adding the transform and inverse transform pairs

$$\sqrt{\frac{2}{\pi}} \int_{0}^{\infty} e^{-ax}\cos x\xi\,dx = \sqrt{\frac{2}{\pi}}\frac{a}{a^2 + \xi^2},$$ (3.229)

$$\sqrt{\frac{2}{\pi}}\int_0^\infty \sqrt{\frac{2}{\pi}}\frac{a}{a^2+x^2}\cos x\xi\, dx = e^{-a\xi}, \tag{3.230}$$

we obtain eigenfunctions

$$\phi(x)=e^{-ax}+\sqrt{\frac{2}{\pi}}\frac{a}{a^2+x^2}. \tag{3.231}$$

This eigenfunction corresponds to the eigenvalue of unity. We could do this with any pair of functions and their transforms, which illustrates the multiple eigenfunctions for our singular integral equation. Also, by subtracting the transform from its function, we obtain eigenvalues of -1.

2. Equations of Convolution Type

An integral equation of the form

$$\frac{1}{\sqrt{2\pi}}\int_{-\infty}^\infty k(x-t)u(t)dt = f(x), \tag{3.232}$$

under the Fourier transform, becomes

$$K(\xi)U(\xi)=F(\xi). \tag{3.233}$$

It may seem that we can solve for u from

$$U(\xi)=M(\xi)F(\xi),\quad M(\xi)=1/K. \tag{3.234}$$

However, the reciprocals of Fourier transforms do not have inverses. If K decays at infinity, M grows at infinity. If this growth is algebraic (as opposed to exponential), we may divide M by a power of ξ, such as ξ^n, and compensate for this by multiplying F by ξ^n. The factor $\xi^{-n}M$ may have an inverse, and the inverse of $\xi^n F$ can be found if the nth derivative of f exists.

As an illustration, consider the equation

$$\frac{1}{\sqrt{2\pi}}\int_{-\infty}^\infty \frac{u(t)dt}{|x-t|^p}=f(x),\quad 0<p<1. \tag{3.235}$$

Taking the Fourier transform

$$\mathcal{F}[|x|^{-p}]U(\xi) = F(\xi). \tag{3.236}$$

Using Eq. (3.109),

$$\sqrt{\frac{2}{\pi}}\Gamma(q)\cos(\pi q/2)|\xi|^{-q}U(\xi) = F(\xi), \quad q \equiv 1 - p. \tag{3.237}$$

$$U = \sqrt{\frac{\pi}{2}}\frac{|\xi|^q}{\Gamma(q)\cos(\pi q/2)}F(\xi). \tag{3.238}$$

Here, ξ^q with $q > 0$ cannot be inverted. By multiplying and dividing by $(-i\xi)$,

$$U = \sqrt{\frac{\pi}{2}}\frac{|\xi|^q}{-i\operatorname{sgn}(\xi)|\xi|\Gamma(q)\cos(\pi q/2)}[-i\xi F(\xi)]$$

$$= i\sqrt{\frac{\pi}{2}}\frac{\operatorname{sgn}(\xi)|\xi|^{-p}}{\Gamma(q)\cos(\pi q/2)}[-i\xi F(\xi)].$$

Assuming the existence of $f'(x)$ and using Eq. (3.110), we invert the two factors to get

$$\mathcal{F}^{-1}[-i\xi F(\xi)] = f'(x), \tag{3.239}$$

$$\mathcal{F}^{-1}[i\operatorname{sgn}(\xi)|\xi|^{-p}] = \sqrt{\frac{\pi}{2}}\frac{\operatorname{sgn}(x)|x|^{-q}}{\Gamma(p)\sin(\pi p/2)}. \tag{3.240}$$

Using convolution,

$$u(x) = \frac{\sqrt{\pi}}{2\sqrt{2}\Gamma(p)\Gamma(q)\sin(\pi p/2)\cos(\pi q/2)}\left[\int_{-\infty}^{x}\frac{f'(t)dt}{(x-t)^q} - \int_{x}^{\infty}\frac{f'(t)dt}{(t-x)^q}\right].$$

$$\tag{3.241}$$

In the special case, $p = 1/2$,

$$u(x) = \frac{1}{\sqrt{2\pi}}\left[\int_{-\infty}^{x}\frac{f'(t)dt}{\sqrt{x-t}} - \int_{x}^{\infty}\frac{f'(t)dt}{\sqrt{t-x}}\right]. \tag{3.242}$$

3. Eddington's Method

In the preceding example, we required the forcing function f to be once-differentiable. If it belongs to the class of infinitely differentiable functions, C^∞, we assume the Fourier transform $U(\xi)$ of the unknown has an expansion

$$U(\xi) = \sum_{n=0}^{\infty} a_n(-i\xi)^n F(\xi), \qquad (3.243)$$

where a_n are unknown. From Eq. (3.233), we get

$$K(\xi) \sum_{n=0}^{\infty} a_n(-i\xi)^n F(\xi) = F(\xi). \qquad (3.244)$$

Thus,

$$\sum_{n=0}^{\infty} a_n(-i\xi)^n = 1/K(\xi). \qquad (3.245)$$

If $1/K(\xi)$ can be expanded in a power series, we can solve for a_n. For a power series expansion, K has to be analytic in ξ around the origin.

Consider the case

$$\frac{1}{\sqrt{2\pi}} \int_{-\infty}^{\infty} e^{-(x-t)^2} u(t)dt = f(x). \qquad (3.246)$$

Taking the Fourier transform,

$$U = \sqrt{2}e^{\xi^2/4}F,$$

$$= \sqrt{2}\left[1 + \left(\frac{\xi^2}{4}\right) + \frac{1}{2!}\left(\frac{\xi^2}{4}\right)^2 + \cdots\right]F. \qquad (3.247)$$

This can be inverted to get

$$u = \sqrt{2}\left[f(x) - \frac{1}{4}f''(x) + \frac{1}{2!}\frac{1}{4^2}f''''(x) + \cdots\right]. \qquad (3.248)$$

4. Evaluation of Integrals

Certain integrals that are difficult to evaluate directly may be obtained using Fourier transforms.

Consider the integral

$$I = \int_0^\infty e^{-(a^2 x^2 + b^2 x^{-2})} dx. \tag{3.249}$$

Taking the Fourier transform with respect to the parameter b,

$$F = \mathcal{F}[I, b \to \xi] = \frac{1}{\sqrt{2}} \int_0^\infty e^{-(a^2 + \xi^2/4)x^2} x\, dx. \tag{3.250}$$

Integrating with respect to x, we have

$$F = \frac{1}{2\sqrt{2}} \frac{1}{a^2 + \xi^2/4} = \frac{1}{\sqrt{2}a} \frac{2a}{4a^2 + \xi^2}. \tag{3.251}$$

Inverting this,

$$I = \frac{\sqrt{\pi}}{2a} e^{-2ab}. \tag{3.252}$$

Another integral that can be evaluated using the Fourier transform is the convolution integral

$$f(x) = \frac{1}{\sqrt{2\pi}} \int_{-\infty}^\infty \frac{dt}{|x-t|^p |t|^q}, \tag{3.253}$$

where $p + q < 1$. Taking the transform and using Eq. (3.109), we find

$$F(\xi) = \frac{2}{\pi} \Gamma(1-p)\Gamma(1-q) \cos\pi[(1-p)/2] \cos\pi[(1-q)/2] |\xi|^{p+q-2}. \tag{3.254}$$

Inverting this, we get

$$f(x) = \sqrt{\frac{2}{\pi}} \frac{\Gamma(1-p)\Gamma(1-q)}{\Gamma(2-p-q)} \frac{\cos\pi[(1-p)/2]\cos\pi[(1-q)/2]}{\cos\pi(2-p-q)/2} |x|^{1-p-q}. \tag{3.255}$$

We may also take advantage of the norm (and inner product) conservation property in the evaluation of integrals. For example,

$$I \equiv \int_{-\infty}^\infty \frac{dx}{(a^2 + x^2)^2} = \|f\|^2, \quad f = \frac{1}{a^2 + x^2}.$$

As $\|f\| = \|F\|$, we have

$$I = \|F\|^2 = \frac{2}{\pi a^2} \int_0^\infty e^{-2a\xi}\, d\xi$$

$$= \frac{1}{\pi a^3}. \tag{3.256}$$

3.12 HILBERT TRANSFORM

In this section, we consider time-dependent real functions $f(t)$, instead of the space-dependent functions we had looted at previously. We further assume $f(t)$ is *causal*, meaning $f(t) = 0$ for $t < 0$. We also assume $f(t)$ is continuous and absolutely integrable. The Cosine and Sine transforms of f may be denoted by

$$X(\tau) = \sqrt{\frac{2}{\pi}} \int_0^\infty f(t)\cos t\tau\, dt, \tag{3.257}$$

$$Y(\tau) = \sqrt{\frac{2}{\pi}} \int_0^\infty f(t)\sin t\tau\, dt. \tag{3.258}$$

The inverse relations are

$$f(t) = \sqrt{\frac{2}{\pi}} \int_0^\infty X(\eta)\cos t\eta\, d\eta \tag{3.259}$$

$$= \sqrt{\frac{2}{\pi}} \int_0^\infty Y(\eta)\sin t\eta\, d\eta. \tag{3.260}$$

Substituting for $f(t)$ from Eq. (3.260) in Eq. (3.257),

$$X(\tau) = \frac{2}{\pi} \int_0^\infty \int_0^\infty Y(\eta)\sin t\eta \cos t\tau\, d\eta dt$$

$$= \frac{1}{\pi} \int_0^\infty \int_0^\infty Y(\eta)[\sin t(\eta + \tau) + \sin t(\eta - \tau)]dt d\eta$$

$$= \frac{1}{\pi} \int_0^\infty Y(\eta)\left[\frac{1}{\eta+\tau} + \frac{1}{\eta-\tau}\right]d\eta,$$

$$= \frac{2}{\pi} \int_0^\infty \frac{\eta Y(\eta)d\eta}{\eta^2 - \tau^2}. \tag{3.261}$$

Similarly, substituting for $f(t)$ and from Eq. (3.259) in Eq. (3.258),

$$Y(\tau) = \frac{2\tau}{\pi} \int_0^\infty \frac{X(\eta)d\eta}{\tau^2 - \eta^2}.$$ (3.262)

Here, X and Y form a transform pair under the Hilbert transform.

Recognizing $X(\tau)$ is an even function of η and $Y(\tau)$ is an odd function,

$$\begin{aligned} X(\tau) &= \frac{1}{\pi} \left[\int_0^\infty \frac{Y(\eta)d\eta}{\eta + \tau} + \int_0^\infty \frac{Y(\eta)d\eta}{\eta - \tau} \right] \\ &= \frac{1}{\pi} \left[\int_{-\infty}^0 \frac{Y(-\eta)d\eta}{-\eta + \tau} + \int_0^\infty \frac{Y(\eta)d\eta}{\eta - \tau} \right] \\ &= \frac{1}{\pi} \int_{-\infty}^\infty \frac{Y(\eta)d\eta}{\eta - \tau}. \end{aligned}$$ (3.263)

Similarly,

$$Y(\tau) = \frac{1}{\pi} \int_{-\infty}^\infty \frac{X(\eta)d\eta}{\tau - \eta}.$$ (3.264)

These represent an alternate form of the Hilbert transform pair

$$Y(t) = \mathcal{H}[X(\tau); \tau \to t] = \frac{1}{\pi} \int_{-\infty}^\infty \frac{X(\tau)d\tau}{t - \tau},$$ (3.265)

$$X(t) = -\mathcal{H}[Y(\tau); \tau \to t] = \frac{1}{\pi} \int_{-\infty}^\infty \frac{Y(\tau)d\tau}{\tau - t}.$$ (3.266)

These relations are also known as the Kramers-Krönig relations.

3.13 CAUCHY PRINCIPAL VALUE

The preceding Hilbert transform pair can also be obtained directly from the Cauchy's theorem,

$$f(z) = \frac{1}{2\pi i} \oint_C \frac{f(s)ds}{s - z},$$ (3.267)

where C is a closed, smooth curve and z is an interior point. When z is on the curve C, if we interpret the integral as an improper integral with a Cauchy principal value, we obtain

$$f(z) = \frac{1}{\pi i} \oint_C^* \frac{f(s)ds}{s-z}, \qquad (3.268)$$

where we follow Tricomi's notation in using the superscript * to indicate the Cauchy principal value. Now, consider the closed curve formed by the real axis, $-\infty < \xi < \infty$, and a semicircle of infinite radius on the upper half plane. Assuming $|f(z)|$ vanishes at infinity,

$$f(x+i0) = \frac{1}{\pi i} \int_{-\infty}^{*\infty} \frac{f(\xi)d\xi}{\xi - x}. \qquad (3.269)$$

If

$$f(x+i0) = u(x,0) + iv(x,0), \qquad (3.270)$$

separating the real and imaginary parts, we find the Hilbert transform pair

$$u(x,0) = -\mathcal{H}[v(\xi,0)] = \frac{1}{\pi} \int_{-\infty}^{*\infty} \frac{v(\xi)d\xi}{\xi - x}, \qquad (3.271)$$

$$v(x,0) = \mathcal{H}[u(\xi,0)] = \frac{1}{\pi} \int_{-\infty}^{*\infty} \frac{u(\xi)d\xi}{x - \xi}. \qquad (3.272)$$

In particular, for $u = \cos \omega x$ and $v = \sin \omega x$,

$$\cos \omega x = \mathcal{H}[\sin \omega \xi], \quad \sin \omega x = -\mathcal{H}[\cos \omega \xi]. \qquad (3.273)$$

In signal processing, a signal $f(t)$ is modulated by multiplying it by $\cos \Omega t$, where Ω is a large number, to get the modulated signal $g(t)$. The function

$$h(t) = g(t) + i\mathcal{H}[g(\tau), \tau \to t], \qquad (3.274)$$

called an **analytic signal** is used for transmission. One advantage of this method is that h can be factored as

$$h(t) = f(t)e^{i\Omega t}, \qquad (3.275)$$

and $f(t) = h(t)e^{-i\Omega t}$ can be reconstructed using fewer sampling points than the modulated signal. A theorem of Nyquist which states that to reconstruct a signal with frequency N cycles per second, we need to sample at least $2N$ points.

3.14 HILBERT TRANSFORM ON A UNIT CIRCLE

Another form of the Hilbert transform for periodic functions in $(0, 2\pi)$ can be obtained from the Cauchy integral

$$f(z) = \frac{1}{2\pi} \oint \frac{f(s)ds}{s - z}, \qquad (3.276)$$

where the integral is on a unit circle C with $s = e^{i\phi}$ a point on the circle and $z = re^{i\theta}$ a point inside the circle. When z is on the circle, we have

$$f(e^{i\theta}) = \frac{1}{\pi} \oint^* \frac{f(s)ds}{s - z}, \qquad |z| = 1,$$

$$= \frac{1}{\pi} \oint^* \frac{f(e^{i\phi})e^{i\phi}id\phi}{e^{i\phi} - e^{i\theta}}$$

$$= \frac{1}{\pi} \oint^* \frac{f(e^{i\phi})id\phi}{1 - e^{i(\theta - \phi)}}$$

$$= \frac{1}{\pi} \oint^* \frac{f(e^{i\phi})e^{-i(\theta - \phi)/2}id\phi}{e^{-i(\theta - \phi)/2} - e^{i(\theta - \phi)/2}}$$

$$= \frac{1}{2\pi} \oint^* f(e^{i\phi})[1 + i\cot(\theta - \phi)/2]d\phi. \qquad (3.277)$$

Let

$$f(e^{i\theta}) = u(\theta) + iv(\theta). \qquad (3.278)$$

Then

$$u(\theta) = \frac{1}{2\pi} \oint^* v(\phi)\cot[(\theta - \phi)/2] + \frac{1}{2\pi} \oint u(\phi)d\phi, \qquad (3.279)$$

$$v(\theta) = \frac{1}{2\pi} \oint^* v(\phi)\cot[(\phi - \theta)/2] + \frac{1}{2\pi} \oint v(\phi)d\phi, \qquad (3.280)$$

For a historical development of the two forms of the Hilbert transforms, the two-volume work on Hilbert transforms by King (2009) is recommended.

3.15 FINITE HILBERT TRANSFORM

Integral equations of the form

$$\frac{1}{\pi} \int_{-1}^{*1} \frac{u(\xi)d\xi}{x-\xi} = v(x), \quad -1 < x < 1, \qquad (3.281)$$

where $u(x)$ has to be found, appear in thin airfoil theory and in elastic contact and crack problems (see Milne-Thomson, 1958, Gladwell, 1980). In thin airfoil theory, circulation around an idealized 2D airfoil is created using distributed vortices of unknown strength along its chord, and the condition that the incompressible fluid velocity has to be tangent to the airfoil profile is used to solve for the unknown. In elasticity, an unknown pressure distribution on the contact line or crack length is related to the known displacement caused by the pressure. The finite Hilbert transform also has applications in image processing.

We may define the finite Hilbert transform as

$$T[v] \equiv \frac{1}{\pi} \int_{-1}^{*1} \frac{v(\xi)d\xi}{x-\xi}. \qquad (3.282)$$

Note the kernel has the form $1/(x-\xi)$ unlike the regular Hilbert transform. This can be related to a line integral in the complex plane in two ways: the Cauchy integral representation of analytic functions and the Plemelj formulas.

3.15.1 Cauchy Integral

As before, consider a unit circle C in the complex z-plane described parametrically by $s = e^{i\phi}, -\pi < \phi < \pi$. Using the Cauchy integral theorem, we define an analytic function inside C by

$$F(z) = \frac{1}{2\pi i} \oint_C \frac{F(s)ds}{s-z}, \quad |z| < 1. \qquad (3.283)$$

When z is on the contour C, the integral has to be evaluated as the Cauchy principal value, and

$$F(z) = \frac{1}{\pi i} \oint_C^* \frac{F(s)ds}{s-z}, \quad |z| = 1. \tag{3.284}$$

If we replace z by $\bar{z}$,

$$F(\bar{z}) = \frac{1}{\pi i} \oint_C^* \frac{F(s)ds}{s-\bar{z}}, \quad |\bar{z}| = 1. \tag{3.285}$$

By adding and subtracting Eqs. (3.284) and (3.285), we find

$$F(z) \pm F(\bar{z}) = \frac{1}{\pi i} \oint_C^* F(s) \left[\frac{1}{s-z} \pm \frac{1}{s-\bar{z}} \right] ds. \tag{3.286}$$

With

$$z = e^{i\theta}, \quad s = e^{i\phi}, \quad ds = isd\phi,$$

we have

$$\left[\frac{1}{s-z} + \frac{1}{s-\bar{z}} \right] isd\phi = \left[i - \frac{\sin\phi}{\cos\phi - \cos\theta} \right] d\phi, \tag{3.287}$$

$$\left[\frac{1}{s-z} - \frac{1}{s-\bar{z}} \right] isd\phi = -\frac{\sin\theta}{\cos\phi - \cos\theta}. \tag{3.288}$$

Using the notation

$$F(e^{i\theta}) = F(\theta), \tag{3.289}$$

Eq. (3.286) for the two cases, plus and minus, becomes

$$F(\theta) + F(-\theta) = \frac{1}{\pi} \oint_C^* F(\phi) \left[1 + i\frac{\sin\phi}{\cos\phi - \cos\theta} \right] d\phi, \tag{3.290}$$

$$F(\theta) - F(-\theta) = \frac{i}{\pi} \oint_C^* F(\phi)\frac{\sin\theta}{\cos\phi - \cos\theta} d\phi. \tag{3.291}$$

For an arbitrary function $G(\phi)$, the integrals around the circle can be split into two integrals as

$$\int_{-\pi}^{\pi} G(\phi)d\phi = \int_0^{\pi} G(\phi)d\phi + \int_{-\pi}^0 G(\phi)d\phi = \int_0^{\pi} [G(\phi) + G(-\phi)]d\phi.$$

Introducing

$$f(\theta) \equiv \frac{1}{2}[F(\theta) + F(-\theta)], \tag{3.292}$$

$$ig(\theta) \equiv \frac{1}{2}[F(\theta) - F(-\theta)], \tag{3.293}$$

which amounts to the relation $F(\theta) = f(\theta) + ig(\theta)$, the preceding equations can be expressed as

$$f(\theta) = \frac{1}{\pi} \left\{ \int_0^{*\pi} f(\phi)d\phi - \int_0^{*\pi} \frac{g(\phi)\sin\phi \, d\phi}{\cos\phi - \cos\theta} \right\}, \tag{3.294}$$

$$g(\theta) = \frac{1}{\pi} \int_0^{*\pi} \frac{f(\phi)\sin\theta \, d\phi}{\cos\theta - \cos\phi}. \tag{3.295}$$

Here, f and g form a finite Hilbert transform pair in terms of trigonometric kernels. This is the common form found in thin airfoil theory.

We note that for a given f, its finite Hilbert transform g is unique, whereas the inverse of g is not unique because of the added term – the average value of f between 0 and π, on the right-hand side of Eq. (3.294).

Using

$$\cos\theta = x, \quad \cos\phi = \xi, \tag{3.296}$$

$$\frac{f(\cos^{-1}x)}{\sqrt{1-x^2}} \Rightarrow u(x), \quad \frac{g(\cos^{-1}x)}{\sqrt{1-x^2}} \Rightarrow v(x), \tag{3.297}$$

in Eqs. (3.295) and (3.294), we get

$$v(x) = \frac{1}{\pi} \int_{-1}^{*1} \frac{u(\xi)d\xi}{x - \xi}, \tag{3.298}$$

$$\sqrt{1-x^2}\,u(x) = \frac{1}{\pi} \int_{-1}^{*1} \frac{\sqrt{1-\xi^2}\,v(\xi)d\xi}{\xi - x} + \frac{1}{\pi} \int_{-1}^{1} u(\xi)d\xi. \tag{3.299}$$

We have the finite Hilbert transform operator

$$T[u, x \to \xi] = v(\xi) \equiv \frac{1}{\pi} \int_{-1}^{1} \frac{u(x)dx}{\xi - x}, \tag{3.300}$$

and its non-unique inverse

$$T^{-1}[v, \xi \rightarrow x] = u(x) \equiv \frac{1}{\pi} \int_{-1}^{*1} \sqrt{\frac{1-\xi^2}{1-x^2}} \frac{v(\xi)d\xi}{\xi - x} + \frac{C}{\sqrt{1-x^2}}, \quad (3.301)$$

where the arbitrary constant

$$C = \frac{1}{\pi} \int_{-1}^{1} u(\xi)d\xi. \quad (3.302)$$

In thin airfoil theory, the physics of the flow requires $u(-1)$ to be finite. With this condition, we get a unique solution in which

$$C = -\frac{1}{\pi} \int_{0}^{\pi} \frac{\sqrt{1-\xi^2}v(\xi)d\xi}{\xi + 1}. \quad (3.303)$$

3.15.2 Plemelj Formulas

The Plemelj formulas deserve consideration on their own merit due to their use in solving singular integral equations arising in the plane theory of elasticity. Here, we derive them to illustrate certain properties of the finite Hilbert transform. Consider a line integral,

$$F(z) = \frac{1}{2\pi i} \int_{C} \frac{f(s)ds}{s - z}, \quad (3.304)$$

from A to B over the curve shown in Fig. 3.11. Unlike our previous Cauchy integrals over a closed curve, this is over an open curve. As we traverse from A to B, the domain close to the curve at left is referred to as the "+"-side and that to the right is the "−"-side. This convention is in agreement with what we have for closed curves. The function F is analytic in the entire complex domain except on the curve itself. When z approaches the curve from the "+"-side, we remove a piece of the curve CD of length 2ϵ and write

$$F_+(z) = \lim_{\epsilon \rightarrow 0} \frac{1}{2\pi i} \left\{ \int_{AC+DB} \frac{f(s)ds}{s - z} + \int_{C'} \frac{f(s)ds}{s - z} \right\}, \quad (3.305)$$

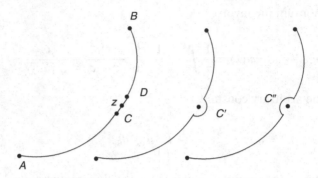

Figure 3.11. Indented contours for Cauchy principal value.

where C' is a semicircle of radius ϵ centered at z (as shown in the second sketch). In the limit of $\epsilon \to 0$, the first integral becomes the Cauchy principal value,

$$F_P(z) = \lim_{\epsilon \to 0} \frac{1}{2\pi i} \int_{AC+DB} \frac{f(s)ds}{s-z}, \tag{3.306}$$

and the second becomes $f(z)/2$. Thus,

$$F_+(z) = F_P(z) + \frac{1}{2}f(z). \tag{3.307}$$

Similarly, when we approach the curve from the "$-$"-side, we take the semi-circle, C'', to the left, to get

$$F_-(z) = F_P(z) - \frac{1}{2}f(z). \tag{3.308}$$

The difference in the signs in front of $f(z)$ in the two relations can be attributed to an angle change from $-\pi$ to 0 on C' and from π to 0 on C''. By adding and subtracting these two relations, we find the Plemelj formulas

$$F_P = \frac{1}{2}[F_+(z) + F_-(z)], \quad f(z) = \frac{1}{2}[F_+(z) - F_-(z)]. \tag{3.309}$$

To relate these results to the finite Hilbert transform, let us define our curve as the straight line $y = 0$ between $x = -1$ and $x = 1$. Since $F(z)$ is continuous in $-\infty < x < -1$ and $1 < x < \infty$, from the Plemelj formulas

$$f(x) = 0, \quad F_P(x) = F(x), \quad y = 0, \quad 1 < |x| < \infty, \tag{3.310}$$

$$F_+(x) - F_-(x) = 2f(x), \quad F_+(x) + F_-(x) = \frac{1}{\pi i} \int_{-1}^{*1} \frac{f(\xi)d\xi}{\xi - x}. \tag{3.311}$$

From our previous calculation with the unit circle where $F(\theta)$ and $F(-\theta)$ are complex conjugates, we choose

$$F_+(x) = U(x) + iV(x), \quad F_-(x) = U(x) - iV(x), \tag{3.312}$$

and rewrite the Plemelj formulas as

$$f(x) = iV(x), \quad U(x) = \frac{1}{2\pi} \int_{-1}^{*1} \frac{V(\xi)d\xi}{\xi - x}. \tag{3.313}$$

From the finite Hilbert transform and its inverse given by Eqs. (3.300) and (3.301), we notice (U, V) and (u, v) are related through the Jacobian of the transformation $x = \cos\theta$. In general, we may transform the unit circle $|z| = 1$ to an ellipse in $\zeta = \xi + i\eta$ plane using the conformal map

$$\zeta = \frac{1}{2}\left[z + \frac{1}{z}\right] + \frac{\epsilon}{2}\left[z - \frac{1}{z}\right]. \tag{3.314}$$

When the parameter ϵ decreases from 1 to 0, the unit circle maps into an ellipse and ultimately ends up as a slit between -1 and 1.

3.16 COMPLEX FOURIER TRANSFORM

The variables x and ξ in the Fourier transform pair, $f(x)$ and $F(\xi)$, are real. When one or both of these are allowed to be complex, we have the complex Fourier transform. Under complex transforms, we have a larger class of functions that are transformable.

Consider the function

$$f(x) = \begin{cases} e^{ax}, & x \geq 0, \\ 0, & x < 0, \end{cases} \tag{3.315}$$

where $a > 0$. As this function is not absolutely integrable, we do not have a regular Fourier transform of $f(x)$. However, we may define

$$g(x) = f(x)e^{-\eta x}, \qquad (3.316)$$

which is absolutely integrable if $\eta > a$. Taking its Fourier transform,

$$G(\xi) = \frac{1}{\sqrt{2\pi}} \int_{-\infty}^{\infty} g(x)e^{i\xi x} dx$$

$$= \frac{1}{\sqrt{2\pi}} \int_{0}^{\infty} e^{(a-\eta+i\xi)x} dx$$

$$= \frac{1}{\sqrt{2\pi}} \int_{0}^{\infty} e^{ax} e^{i(\xi+i\eta)x} dx$$

$$= \frac{1}{\sqrt{2\pi}} \int_{-\infty}^{\infty} f(x)e^{i\zeta x} dx = F(\zeta), \qquad (3.317)$$

with the constraint

$$\text{Im}(\zeta) = \eta > a. \qquad (3.318)$$

In the complex ζ–plane, the region $\eta \leq a$ contains all the singularities of the complex function $F(\zeta)$. Or equivalently, $F(\zeta)$ is analytic above the line $\eta = a$. We have $F(\zeta)$ as the complex Fourier transform of $f(x)$.

To obtain an inversion formula, we, again, work with $g(x)$. We have

$$g(x) = f(x)e^{-\eta x} = \frac{1}{\sqrt{2\pi}} \int_{-\infty}^{\infty} G(\xi)e^{-i\xi x} d\xi,$$

$$f(x) = \frac{1}{\sqrt{2\pi}} \int_{-\infty}^{\infty} G(\xi)e^{(\eta-i\xi)x} d\xi$$

$$= \frac{1}{\sqrt{2\pi}} \int_{C} F(\zeta)e^{-i\zeta x} d\zeta, \qquad (3.319)$$

where the line C is defined by $\eta = $ constant $(> a)$. Figure 3.12 shows the analytic domain for the complex function $F(\zeta)$ and the inversion contour $C : \eta > a$.

Next, we consider a function $f(x)$ defined as zero for positive values of x and bounded by e^{-bx} for negative values of x. Suppose we multiply this function by $e^{-\eta x}$ to make it absolutely integrable. Then η has to

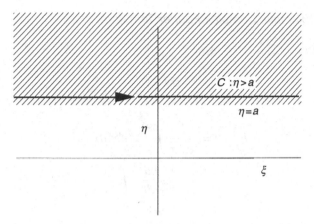

Figure 3.12. Analytic domain for $F(\zeta)$ and the integration contour C for inversion.

satisfy $\eta < -b$. Thus, the Fourier transform $F(\zeta)$ of $f(x)$ will be analytic below the line $\eta = -b$ and the inversion contour can be any line below $\eta = -b$.

In the normal case of functions defined on $-\infty < x < \infty$, we may define

$$f_+(x) = \begin{cases} f(x), & x \geq 0 \\ 0, & x < 0 \end{cases}, \qquad f_-(x) = \begin{cases} 0, & x \geq 0 \\ f(x), & x < 0 \end{cases}. \qquad (3.320)$$

Then

$$f(x) = f_+(x) + f_-(x), \qquad (3.321)$$

and each of the functions on the right-hand side has its own complex Fourier transform. If we use $F_+(\zeta)$ and $F_-(\zeta)$ to denote these transforms, the inversion contours must lie entirely in the analytic domains of these functions. We may use the complex transform to include functions that are not absolutely integrable as long as they are bounded by an exponentially growing function.

In this light, a function $k(x)$, which allows regular Fourier transform, may behave like e^{-ax} for $x \to \infty$ and e^{bx} for $x \to -\infty$; its complex Fourier transform is analytic above the line $\eta = -a$ and below $\eta = b$.

3.16.1 Example: Complex Fourier Transform of x^2

By writing

$$x^2 = f_+ + f_-, \tag{3.322}$$

we have

$$F_+(\zeta) = \frac{1}{\sqrt{2\pi}} \int_0^\infty x^2 e^{i\zeta x} dx$$

$$= -\frac{1}{\sqrt{2\pi}} \frac{d^2}{d\zeta^2} \int_0^\infty e^{i\zeta x} dx$$

$$= \frac{1}{\sqrt{2\pi}} \frac{d^2}{d\zeta^2} \frac{1}{i\zeta}$$

$$= -\frac{2i}{\sqrt{2\pi}} \frac{1}{\zeta^3}, \tag{3.323}$$

$$F_-(\zeta) = \frac{1}{\sqrt{2\pi}} \int_{-\infty}^0 x^2 e^{i\zeta x} dx$$

$$= \frac{2i}{\sqrt{2\pi}} \frac{1}{\zeta^3}. \tag{3.324}$$

Both of these functions have poles at $\zeta = 0$. The inversion contour for F_+ is a line above the pole, and that for F_- is a line below the pole.

3.16.2 Example: Complex Fourier Transform of $e^{|x|}$

Again, by writing

$$e^{|x|} = f_+ + f_-, \tag{3.325}$$

$$F_+(\zeta) = \frac{1}{\sqrt{2\pi}} \int_0^\infty e^{(1+i\zeta)x} dx$$

$$= -\frac{1}{\sqrt{2\pi}} \frac{1}{1+i\zeta}, \tag{3.326}$$

$$F_-(\zeta) = \frac{1}{\sqrt{2\pi}} \int_{-\infty}^0 e^{(-1+i\zeta)x} dx$$

$$= \frac{1}{\sqrt{2\pi}} \frac{1}{-1+i\zeta}. \tag{3.327}$$

Here, F_+ has a pole at $\zeta = i$, and it is analytic above the line $\eta = 1$. F_- has a pole at $\zeta = -i$, and it is analytic below the line $\eta = -1$.

3.17 WIENER-HOPF METHOD

The Wiener-Hopf method uses the idea of partitioning a complex Fourier transform into F_+ and F_- and inverting them using separate contour integrals. The function F_+ must be analytic above a certain line, $\text{Im}(\zeta) = a$, and F_- must be analytic below a certain line, $\text{Im}(\zeta) = b$. This method has found many applications involving mixed boundary value problems, which have the dependent function prescribed on part of the boundary and its derivative prescribed on the remaining part of the boundary. In the following, we discuss some classical examples from Morse and Feshbach (1953).

3.17.1 Example: Integral Equation

Consider the integral equation

$$u(x) - \int_{-\infty}^{\infty} e^{-|x-t|} u(t)dt = x^2. \tag{3.328}$$

When this is put in the form of a convolution integral, the kernel is

$$k(x) = \sqrt{2\pi}e^{-|x|}. \tag{3.329}$$

We observe that the forcing function allows complex Fourier transforms

$$F_+ = -\frac{2i}{\sqrt{2\pi}} \frac{1}{\zeta^3}, \quad F_- = \frac{2i}{\sqrt{2\pi}} \frac{1}{\zeta^3}. \tag{3.330}$$

We assume the unknown u also is made of two parts, u_+ and u_-, with

$$u_+ = \frac{1}{\sqrt{2\pi}} \int_{C_+} U_+ e^{-i\zeta x}(\zeta)d\zeta, \quad u_- = \frac{1}{\sqrt{2\pi}} \int_{C_-} U_- e^{-i\zeta x}(\zeta)d\zeta, \tag{3.331}$$

where the two lines, C_+ and C_-, have to be selected. The Fourier transform of the kernel is

$$K(\zeta) = \frac{2}{\zeta^2 + 1}, \tag{3.332}$$

which is analytic in the strip: $-1 < \eta < 1$. The inversion contour C_+ has to be chosen to run through the analytic region of K; at the same time, U_+ must be analytic above this line. Similarly, C_- also has to be in the analytic region of K with U_- analytic below it.

Substituting the inverse transforms into the integral equation

$$\int_{C_+} \left[\left(1 - \frac{2}{\zeta^2+1}\right) U_+ - F_+ \right] e^{-i\zeta x} d\zeta$$
$$+ \int_{C_-} \left[\left(1 - \frac{2}{\zeta^2+1}\right) U_- - F_- \right] e^{-i\zeta x} d\zeta = 0. \tag{3.333}$$

This may be simplified to get

$$\int_{C_+} \left[\frac{\zeta^2-1}{\zeta^2+1} U_+ - F_+ \right] e^{-i\zeta x} d\zeta + \int_{C_-} \left[\frac{\zeta^2-1}{\zeta^2+1} U_- - F_- \right] e^{-i\zeta x} d\zeta = 0. \tag{3.334}$$

Observing from Fig. 3.13 that a closed contour C can be formed by going along C_- and returning from infinity along C_+ (opposite of the direction of the arrow) and again rejoining with C_-. For closed curves, we have from Goursat's theorem

$$\oint_C H(\zeta) e^{-i\zeta x} d\zeta = 0, \tag{3.335}$$

for any function $H(\zeta)$, which is analytic inside the contour. Comparing this with Eq. (3.334), we conclude

$$\frac{\zeta^2-1}{\zeta^2+1} U_+ - F_+ = -H, \tag{3.336}$$

$$\frac{\zeta^2-1}{\zeta^2+1} U_- - F_- = H. \tag{3.337}$$

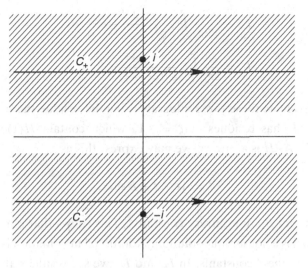

Figure 3.13. Inversion contours C_+ and C_- in the ζ-plane.

Solving for the unknown transforms,

$$U_+ = \frac{\zeta^2 + 1}{\zeta^2 - 1}(F_+ - H), \qquad (3.338)$$

$$U_- = \frac{\zeta^2 + 1}{\zeta^2 - 1}(F_+ + H). \qquad (3.339)$$

These functions have singularities at $\zeta = 0$ (from F_+ and F_-) and at $\zeta = \pm 1$. The inversion contour C_+ can be selected with $\eta > 0$ for it to be above the singular points. However, the kernel K is only analytic between i and $-i$. This restricts C_+ to be below $\eta = 1$. Similarly, C_- is in the strip: $-1 < \eta < 0$.

Using the inversion formula, we get

$$u(x) = I + I_+ + I_-, \qquad (3.340)$$

where

$$I = \frac{1}{\sqrt{2\pi}} \oint_C \frac{\zeta^2 + 1}{\zeta^2 - 1} H(\zeta) e^{-i\zeta x} d\zeta, \qquad (3.341)$$

$$I_+ = -\frac{2i}{2\pi} \int_{C_+} \frac{\zeta^2+1}{\zeta^3(\zeta^2-1)} e^{-i\zeta x} d\zeta, \tag{3.342}$$

$$I_- = \frac{2i}{2\pi} \int_{C_-} \frac{\zeta^2+1}{\zeta^3(\zeta^2-1)} e^{-i\zeta x} d\zeta. \tag{3.343}$$

We evaluate these three integrals using the residue theorem. The integral, I, has residues at $\zeta = \pm 1$, which contain $H(1)e^{-ix}$ and $H(-1)e^{ix}$. As H is arbitrary, we may express this as

$$I = A\cos x + B\sin x, \tag{3.344}$$

where A and B are arbitrary constants. This shows that our integral equation has a non-unique solution. Additional information is needed to evaluate these constants. In I_+ and I_-, we see residues at $\zeta = \pm 1$. As we already have the constants, A and B, contributions from these residues can be omitted. The lines C_+ and C_- can be made part of a closed contour by including semicircles extending to infinity. When x is positive, $e^{-i\zeta x}$ goes to zero as $\eta \to -\infty$, the semicircles are in the lower complex plane (see Fig. 3.14). When x is negative, it is in the

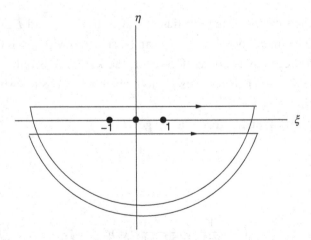

Figure 3.14. Closed contours for evaluating the integrals when $x > 0$.

upper complex plane. Using the residue at $\zeta = 0$, we get

$$I_+ = (4 - x^2)h(x), \quad I_- = (4 - x^2)h(-x), \tag{3.345}$$

where $h(x)$ is the Heaviside step function. Finally, the solution of the original integral equation is

$$u(x) = A\cos x + B\sin x + 4 - x^2. \tag{3.346}$$

We may verify this solution as follows: Differentiate the integral equation twice to convert it to the differential equation

$$u'' + u = 2 - x^2, \tag{3.347}$$

which is satisfied by the given solution. Of course, this is a special case; in general, integral equations cannot be converted to differential equations.

3.17.2 Example: Factoring the Kernel

The Wiener-Hopf method originated from an attempt to solve one-sided integral equations of the form

$$u(x) = \lambda \int_0^\infty k(x - t)u(t)dt, \tag{3.348}$$

where $-\infty < x < \infty$. This equation is usually supplemented with conditions on the growth of u as $|x| \to \infty$.

As in the previous example, we express u using two functions: u_+ and u_- defined as

$$u_+ = \begin{cases} u(x), & x > 0 \\ 0, & x < 0 \end{cases}, \quad u_- = \begin{cases} 0, & x > 0 \\ u(x), & x < 0 \end{cases}. \tag{3.349}$$

Then, we have

$$u_+ + u_- = \lambda \int_{\infty}^\infty k(x - t)u_+(t)dt. \tag{3.350}$$

Assuming the complex Fourier transform of k, namely, $K(\zeta)$, is analytic in a strip, $-a < \mathrm{Im}(\zeta) < b$ (which implies that k behaves as the function e^{-ax} for large positive x and as e^{bx} for large negative x), from the integral equation we can assess the behavior of $u(x)$ for large values of $|x|$. As $x \to \infty$, if $u_+(x) \sim e^{-cx}$, for the integral of $k(x-t)u_+(t) \sim e^{-a(x-t)-ct}$ to converge, $c > a$. As $x \to -\infty$,

$$u_-(x) \sim \int e^{b(x-t)} e^{-ct} dt, \quad u_-(x) \sim e^{bx}. \qquad (3.351)$$

We also require $b + c > 0$ or $c > -b$. Also, $U_+(\zeta)$ is analytic above $\eta = -c$, and $U_-(\zeta)$ is analytic below $\eta = b$. Thus, there is a patch between $\eta = -c$ and $\eta = b$ where U_+ and U_- are analytic. This is shown in Fig. 3.15. These two functions are said to be the analytic continuation of each other. The Fourier transform of the integral equation is

$$(1 - \sqrt{2\pi}\lambda K)U_+ + U_- = 0, \quad \text{or} \quad (1 - \sqrt{2\pi}\lambda K)U_+ = -U_-. \ (3.352)$$

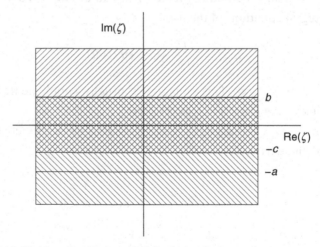

Figure 3.15. The function U_+ is analytic in the hatched area above $\eta = -c$ and U_- is analytic in the hatched area below $\eta = b$. The kernel K is analytic in the strip between b and $-a$.

We hope by factoring

$$(1 - \sqrt{2\pi}\lambda K) = \frac{K_+}{K_-}, \tag{3.353}$$

the resulting equations,

$$K_+ U_+ = -K_- U_-, \tag{3.354}$$

can be solved as

$$U_+ = H/K_+, \quad U_- = -H/K_-, \tag{3.355}$$

where H is analytic between $\eta = -c$ and $\eta = b$, U_+ is analytic above $\eta = -c$, and U_- is analytic below $\eta = b$.

Then

$$u_+ = \frac{1}{\sqrt{2\pi}} \int_C e^{-i\zeta x} \frac{H}{K_+} d\zeta, \quad u_- = -\frac{1}{\sqrt{2\pi}} \int_C e^{-i\zeta x} \frac{H}{K_-} d\zeta. \tag{3.356}$$

For example, if

$$k(x) = e^{-|x|}, \quad K = \sqrt{\frac{2}{\pi}} \frac{1}{\zeta^2 + 1}. \tag{3.357}$$

Here, $a = 1$ and $b = 1$ as K has poles at $\pm i$ such that

$$(1 - \sqrt{2\pi}\lambda K) = \frac{\zeta^2 + 1 - 2\lambda}{\zeta^2 + 1} = \frac{K_+}{K_-}. \tag{3.358}$$

From the two poles at $\pm i$, we use the pole at $-i$ in K_+ and the one at $+i$ in K_-, by the choice

$$K_+ = \frac{\zeta^2 + 1 - 2\lambda}{\zeta + i}, \quad K_- = \zeta - i. \tag{3.359}$$

Of course, this choice is not unique; we may multiply K_+ by any function and divide K_- by the same function as long as we do not violate the regions of analyticity.

As $U_+(\zeta) \to 0$ as $\xi \to \infty$, we obtain $H(\zeta) = D$, a constant. Then

$$U_+ = D \frac{\zeta + i}{\zeta^2 + 1 - 2\lambda}, \quad U_- = -D/(\zeta - i), \tag{3.360}$$

and from Eq. (3.356), we find

$$u_+ = A\left[\cos\mu x + \frac{\sin\mu x}{\mu}\right], \quad \mu = \sqrt{2\lambda - 1}, \qquad (3.361)$$

$$u_- = Ae^x, \qquad (3.362)$$

where the constant A (which is a multiple of D) is indeterminate for our homogeneous equation.

The definitive reference for the Wiener-Hopf method is the book by Noble (1958). More on this method and on analytic continuation may be found in Morse and Feshback (1953) and Davies (1984).

3.18 DISCRETE FOURIER TRANSFORMS

An analog signal, $x(t)$, which is a continuous function of time is converted to a sequence of digital quantities by sampling the signal at fixed intervals, T. Practical constraints limit the number of samples to a finite number, N. We may denote these N sample values by

$$x_n = x(nT), \quad n = 0,1,\ldots,N-1. \qquad (3.363)$$

Beyond the range $[0,(N-1)T]$, we assume the signal is periodic. The discrete Fourier transform of the sequence $\{x_n\}$ is defined as

$$X(\omega) = \sum_{n=0}^{N-1} x_n e^{-i\omega nT}\, T. \qquad (3.364)$$

Here we employ the electrical engineering convention of using $(-i)$ to take the transform. The infinite integral of the regular Fourier transform has been replaced by a finite sum (which is equivalent to numerical integration using the trapezoidal rule). The angular frequency ω is related to the frequency f (cycles per second) by

$$\omega = 2\pi f. \qquad (3.365)$$

Then

$$X(f) = T \sum_{n=0}^{N-1} x_n e^{-2\pi i f n T}. \tag{3.366}$$

The transform $X(f)$ varies as f increases from 0 to $1/T$, and then it repeats in a periodic manner with period $1/T$ due to the periodic nature of the exponential term. This frequency range may be divided into N discrete values, to get

$$\mathcal{F}_d[x_n, n \to m] = X_m = X\left(\frac{m}{NT}\right) = T \sum_{n=0}^{N-1} x_n e^{-2\pi i m n/N}. \tag{3.367}$$

In practice, we are given a sequence $\{x_n\}$ and the interval T is taken as unity. To obtain the inversion formula, we multiply both sides by $e^{2\pi i m j/N}$ and sum:

$$\sum_{m=0}^{N-1} X_m e^{2\pi i m j/N} = \sum_{j=0}^{N-1} \sum_{m=0}^{N-1} x_n e^{-2\pi i m(n-j)/N}$$

$$= T \sum_{n=0}^{N-1} x_n S_{nj}, \tag{3.368}$$

where

$$S_{nj} = \sum_{m=0}^{N-1} e^{-2\pi i m k/N}, \quad k = n - j. \tag{3.369}$$

When $k = 0$, that is, $n = j$, we get

$$S_{nn} = N, \tag{3.370}$$

and when $k \neq 0$, we use the summation formula for a geometric series,

$$\sum_{m=0}^{N-1} \rho^m = \frac{1 - \rho^N}{1 - \rho}, \quad \rho = e^{2\pi i k/N}, \tag{3.371}$$

to get $S_{nj} = 0$. Thus,

$$x_n = \mathcal{F}_d^{-1}[X_m; m \to n] = \frac{1}{NT} \sum_{m=0}^{N-1} X_m e^{2\pi i m j/N}. \tag{3.372}$$

We note the periodicity of the transform pair

$$x_n = x_{n+N}, \quad X_m = X_{m+N}. \tag{3.373}$$

Because of this periodicity, we may visualize the N values of x_n or X_m stored at N equally spaced locations on a circle, and the summations run around this circle.

Following the approach used in continuous Fourier transforms, we may show the convolution sum of two sequences, $\{x_n\}$ and $\{y_n\}$, defined by

$$(x * y)_k = \sum_n x_n y_{k-n}, \tag{3.374}$$

has the property

$$\mathcal{F}_d[x * y; k \to m] = \sum_k \sum_n x_n y_{n-k} e^{-2\pi i k m/N}$$

$$= \sum_j \sum_n x_n y_j e^{-2\pi i (n+j)m/N}, \quad j = n - k,$$

$$= X_m Y_m. \tag{3.375}$$

Using the inverse transform,

$$(x * y)_k = \frac{1}{N} \sum_m X_m Y_m e^{2\pi i k m/N}. \tag{3.376}$$

In this, if we choose $k = 0$,

$$\sum_n x_n y_{-n} = \frac{1}{N} \sum_m X_m Y_m. \tag{3.377}$$

If we choose

$$y_{-n} = x_n^*, \tag{3.378}$$

which is the complex conjugate of x_n,

$$Y_m = \sum_n y_n e^{-2\pi i n m/N} = \sum_n x_{-n}^* e^{-2\pi i n m/N}$$

$$= \sum_n x_n^* e^{2\pi i n m/N} = X_m^*. \tag{3.379}$$

Thus, from Eq. (3.377), the norms of the sequences defined by

$$\|x\| = \sum_{n=0}^{N-1} x_n x_n^*, \quad \|X\| = \frac{1}{N} \sum_{m=0}^{N-1} X_m X_m^*, \tag{3.380}$$

satisfy the Parseval's formula,

$$\|x\| = \|X\|. \tag{3.381}$$

3.18.1 Fast Fourier Transform

To compute one value X_1, we have to sum N terms involving the products of x_n and $e^{-2\pi inm/N}$. In computational terminology, these constitute N "add-multiply" operations. To compute all values of X_m, we expect N^2 operations. In a historic paper, Cooley and Tukey (1965) showed that, if N has the form 2^p and the computations are done in binary arithmetic, the discrete Fourier transform can be accomplished in $N \log_2 N$ operations. When N is large, there is substantial difference between N^2 and $N \log_2 N$. This algorithm has come to be known as the **Fast Fourier transform**. The reader may consult the original paper of 1965 or Andrews and Shivamoggi (1988) for details. When a real array x_n is transformed, X_ms come out as pairs of complex conjugates. Available software, normally store only the real and imaginary parts – thereby reducing the storage need. When reconstructing the original sequence from its discrete transform, terms involving the complex conjugates have to be added.

SUGGESTED READING

Andrews, L. C., and Shivamoggi, B. K. (1988). *Integral Transforms for Engineers and Applied Mathematicians*, Macmillan.

Cooley, J. W., and Tukey, J. W. (1965). An algorithm for the machine calculation of complex Fourier series, *Math. Comp.*, Vol. 19, pp. 297–301.

Davies, B. (1984). *Integral Transforms and Their Applications*, 2nd ed., Springer–Verlag.

Gladwell, G. M. L. (1980). *Contact Problems in Classical Elasticity*, Sijthoff and Noordhoff.

King, F. (2009). *Hilbert Transforms*, Vol. I& II, Cambridge.

Milne-Thomson, L. M. (1958). *Theoretical Aerodynamics*, Macmillan.

Morse, P. M., and Feshbach, H. (1953). *Methods of Applied Mathematics*, Part I, McGraw-Hill.

Noble, B. (1958). *Methods based on the Wiener-Hopf technique for the solution of partial differential equations*, Pergamon Press.

Sneddon, I. N. (1972). *The Use of Integral Transforms*, McGraw–Hill.

EXERCISES

3.1 From the Fourier transform of

$$f(x) = h(1 - |x|),$$

show that

$$\int_0^\infty \frac{\sin x}{x} dx = \frac{\pi}{2}.$$

3.2 Obtain the Cosine and Sine transforms of

$$f = e^{-ax} \cos bx,$$

where a and b are positive constants.

3.3 From the Fourier transform of

$$f(x) = (1 - |x|)h(1 - |x|),$$

compute the integrals

$$I_n = \int_0^\infty \left(\frac{\sin x}{x} \right)^n dx,$$

for $n = 2, 3, 4$.

3.4 Obtain the Fourier transform of

$$f(x) = e^{-ax^2} \cos bx,$$

where a and b are positive constants.

3.5 Compute the Sine transform of

$$f = \frac{e^{-ax}}{x},$$

and find its limit as $a \to 0$.

3.6 Using $f = 1/\sqrt{x}$ in the cosine and sine forms of the Fourier integral theorem, show that

$$\int_0^\infty \frac{\cos x}{\sqrt{x}} dx = \int_0^\infty \frac{\sin x}{\sqrt{x}} dx = \sqrt{\frac{\pi}{2}}$$

and

$$\mathcal{F}_c[1/\sqrt{x}] = \mathcal{F}_s[1/\sqrt{x}] = 1/\sqrt{\xi}.$$

3.7 Invert the following

(a) $F(\xi) = \dfrac{1}{\xi^2 + 4\xi + 13}$,

(b) $F(\xi) = \dfrac{1}{(\xi^2 + 4)^2}$.

3.8 Invert

$$F(\xi) = \frac{e^{-\xi^2}}{\xi^2 + 4},$$

in terms of complementary error functions.

3.9 Invert

$$F(\xi) = \frac{e^{-ia\xi}}{(\xi - 1)^2 + 4}.$$

3.10 For $0 < a < \pi$, compute

$$f(x) = \mathcal{F}^{-1}\left[\frac{\cosh a\xi}{\sinh \pi \xi}\right], \quad g(x) = \mathcal{F}^{-1}\left[\frac{\sinh a\xi}{\sinh \pi \xi}\right].$$

3.11 Evaluate the integral

$$I = \int_{-\infty}^\infty \frac{dx}{(x^2 + a^2)(x^2 + b^2)},$$

by (a) using partial fractions and (b) using the convolution theorem.

3.12 For

$$f(x) = h(1 - |x|),$$

show that $\|f\| = \|F\|$.

3.13 Obtain the Cosine transform of

$$f = \cos\left(\frac{x^2}{2} - \frac{\pi}{8}\right),$$

and show that it is self-reciprocal.

3.14 Show that

$$\frac{1}{2} + \sum_{n=1}^{N} \cos n\theta = \frac{\sin(2N+1)\theta/2}{2\sin\theta/2}.$$

3.15 Using the definition

$$F_c(\xi) = \sqrt{\frac{2}{\pi}} \int_0^\infty f(x) \cos x\xi \, dx,$$

show that

$$\frac{1}{2} F_c(0) + \sum_{n=1}^{N} F_c(n\xi) = \frac{1}{\sqrt{2\pi}} \int_0^\infty f(x) \frac{\sin(2N+1)x\xi/2}{\sin x\xi/2} dx.$$

3.16 Using the Riemann-Lebesgue localization lemma, show that the preceding relation yields the Poisson sum formula

$$\sqrt{\xi} \left[\frac{1}{2} F_c(0) + \sum_{n=1}^{\infty} F_c(n\xi) \right] = \sqrt{x} \left[\frac{1}{2} f(0) + \sum_{n}^{\infty} f(nx) \right],$$

where $x\xi = 2\pi$.

3.17 From the preceding result, show that the Theta function given by the sum

$$\theta(x) = \sum_{-\infty}^{\infty} e^{-n^2 x^2/2}$$

satisfies

$$\theta(x) = \frac{\sqrt{2\pi}}{x} \theta(2\pi/x).$$

3.18 Solve the wave equation for an infinite string

$$\frac{\partial^2 u}{\partial x^2} = \frac{\partial^2 u}{\partial t^2},$$

with the initial conditions

$$u(x,0) = U_0 e^{-x^2}, \quad \frac{\partial u(x,0)}{\partial t} = 0.$$

3.19 Solve the wave equation for an infinite string

$$\frac{\partial^2 u}{\partial x^2} = \frac{\partial^2 u}{\partial t^2},$$

with the initial conditions

$$u(x,0) = \frac{U_0}{x^2 + 1}, \quad \frac{\partial u(x,0)}{\partial t} = 0.$$

3.20 Lateral motion of an infinite beam satisfies

$$EI\frac{\partial^4 v}{\partial x^4} + m\frac{\partial^2 v}{\partial t^2} = 0,$$

where EI is the bending stiffness and m is the mass per unit length. If the initial conditions are

$$v(x,0) = V_0 e^{-x^2}, \quad \frac{\partial v(x,0)}{\partial t} = 0,$$

and $v \to 0$ as $|x| \to \infty$, obtain $v(x,t)$.

3.21 Consider the motion of a concentrated force on an elastically supported taut string. The deflection u satisfies

$$\frac{\partial^2 u}{\partial x^2} + u = \frac{\partial^2 u}{\partial t^2} - \delta(x - ct),$$

with $u \to 0$ as $|x| \to \infty$. Assume the speed of the force, $c > 1$. We are interested in the steady-state distribution of the deflection. Introducing the new variables,

$$\xi = x - ct, \quad \tau = t,$$

rewrite the governing equation. For the steady state, let $\partial u/\partial \tau$ and $\partial^2 u/\partial \tau^2$ go to zero. Obtain the solution that is continuous at $\xi = 0$.

3.22 The transient heat conduction in a 1D infinite bar satisfies

$$\frac{\partial^2 u}{\partial x^2} = \frac{\partial u}{\partial t},$$

with the initial condition

$$u(x,0) = T_0 e^{-|x|}.$$

Obtain the temperature $u(x,t)$.

3.23 The transient heat conduction in a 1D infinite bar with heat generation satisfies

$$\frac{\partial^2 u}{\partial x^2} = \frac{\partial u}{\partial t} + \frac{1}{x^2+1},$$

with the initial condition $u(x,0) = T_0$. Obtain $u(x,t)$ in the form of a convolution integral.

3.24 The transient heat conduction in a 2D infinite plate satisfies

$$\frac{\partial^2 u}{\partial x^2} + \frac{\partial^2 u}{\partial y^2} = \frac{\partial u}{\partial t},$$

with

$$u(x,y,0) = U_0 e^{-(x^2+y^2)}.$$

Obtain $u(x,y,t)$.

3.25 The transient heat conduction in a 2D infinite plate satisfies

$$\frac{\partial^2 u}{\partial x^2} + \frac{\partial^2 u}{\partial y^2} = \frac{\partial u}{\partial t},$$

with

$$u(x,y,0) = U_0 h(1 - |x|)h(1 - |y|).$$

Obtain $u(x,y,t)$.

3.26 A semi-infinite rod has an initial temperature distribution

$$u(x,0) = T_0 x e^{-x^2},$$

and the transient heat conduction is governed by

$$\frac{\partial^2 u}{\partial x^2} = \frac{\partial u}{\partial t}.$$

If the boundary temperature is $u(0,t) = 0$, obtain the temperature $u(x,t)$.

3.27 In the preceding problem, if we replace the condition at $x = 0$ by

$$\frac{\partial u(0,t)}{\partial x} - hu(0,t) = 0,$$

where h is a constant, obtain the solution using the mixed trigonometric transform.

3.28 Solve the integral equation

$$u(x) = xe^{-|x|} - \int_{-\infty}^{\infty} u(t)e^{-|x-t|}dt.$$

3.29 Find the solution of

$$u(x) = e^{-2|x|} - \int_{-\infty}^{\infty} u(t)e^{-|x-t|}dt.$$

3.30 Find the solution of

$$u(x) = |x|^{-p} - \int_{-\infty}^{\infty} u(t)e^{-|x-t|}dt,$$

where p is a real constant: $0 < p < 1$.

3.31 Solve

$$u(x) = xe^{-x^2} - \frac{1}{\sqrt{\pi}}\int_{-\infty}^{\infty} e^{-(x-t)^2} u(t)dt,$$

in the form of an infinite series.

3.32 Solve

$$u(x) = \frac{x}{x^2+4} - \frac{1}{\sqrt{\pi}}\int_{-\infty}^{\infty} e^{-(x-t)^2} u(t)dt,$$

in the form of a series of convolution integrals.

3.33 Solve the system of integral equations

$$u_1(x) = \frac{1}{2} \int_{-\infty}^{\infty} u_2(t) e^{-|x-t|} dt + 2e^{-|x|},$$

$$u_2(x) = \frac{1}{2} \int_{-\infty}^{\infty} u_1(t) e^{-|x-t|} dt - 4e^{-2|x|}.$$

3.34 Show that

$$u = 1/2 \quad \text{and} \quad u = c - x^2$$

are solutions of

$$u(x) \pm \frac{1}{\sqrt{2\pi}} \int_{-\infty}^{\infty} e^{-t^2/2} u(x-t) dt = 1,$$

corresponding to "+" and "−" signs, respectively. Here, c is a constant.

3.35 Obtain all the solutions of the integral equation

$$u(x) - \frac{1}{\sqrt{2\pi}} \int_{-\infty}^{\infty} e^{-t^2/2} u(x-t) dt = 1.$$

3.36 Solve

$$u(x) - \frac{1}{2} \int_{-\infty}^{\infty} e^{-2|x-t|} u(t) dt = e^{|x|}.$$

3.37 Obtain all the solutions of the integral equation

$$u(x) - \frac{\lambda}{\sqrt{2\pi}} \int_{-\infty}^{\infty} e^{-t^2/2} u(x-t) dt = x,$$

with λ being real and positive.

3.38 Solve

$$u(x) - \lambda \int_{-\infty}^{\infty} e^{-|x-t|} u(t) dt = \cos x.$$

3.39 Consider the signal

$$f(t) = 3\cos 3t + 4\sin 2t,$$

which is modulated to get

$$g(t) = f(t)\cos 100t.$$

Show that the analytical signal

$$h(t) = g(t) + i\mathcal{H}[g(\tau), \tau \to t]$$

allows the factoring

$$h(t) = f(t)e^{i100t}.$$

3.40 Show that the finite Hilbert transform of $(1-x^2)^{1/2}$ is x and that of $(1-x^2)^{-1/2}$ is 0. Use the change of variable,

$$x = \frac{1-t^2}{1+t^2},$$

to simplify the integrals.

3.41 Show that the finite Hilbert transform of an even function in $(-1, 1)$ is odd and that of an odd function is even.

3.42 Show that

$$\frac{1}{\pi}\int_0^{*\pi} \frac{\cos n\phi\, d\phi}{\cos\theta - \cos\phi} = -\frac{\sin n\theta}{\sin\theta}.$$

In Eq. (3.295), assuming Fourier series

$$f(\theta) = \sum_{n=0} A_n \cos n\theta, \quad g(\theta) = \sum_{n=1} B_n \sin n\theta,$$

show that $B_n = -A_n$ $(n = 1, 2, \ldots)$. In airfoil theory, B_n are known and A_n are to be found.

4

LAPLACE TRANSFORMS

As the name implies, this integral transform was introduced by Laplace during his studies of probability. The English mathematician and engineer Oliver Heaviside had extensively used the methods of Laplace transforms in the name of operational calculus in solving linear differential equations arising from electrical networks. The Laplace transform, in its common form, applies to functions that are causal. This limitation, compared to the Fourier transform, is more than compensated by the fact that the Laplace transform can be applied to exponentially growing functions. These two features make the Laplace transform ideal for time-dependent functions, which are zero for $t < 0$ and bounded by an exponentially growing function. In this chapter, we assume that, unless otherwise defined, all functions of time are zero for negative values of their arguments.

The Laplace transform of $f(t)$ is defined as

$$\mathcal{L}[f(t), t \to p] = \bar{f}(p) \equiv \int_0^\infty f(t) e^{-pt} dt. \tag{4.1}$$

We assume $f(t)$ is piecewise continuous and satisfies the generalized integrability condition,

$$\left| \int_0^\infty f(t) e^{-\gamma t} dt \right| < M, \tag{4.2}$$

where M and γ are real. The defining integral, Eq. (4.1), in conjunction with the integrability condition implies $\bar{f}(p)$ has no singularities in the

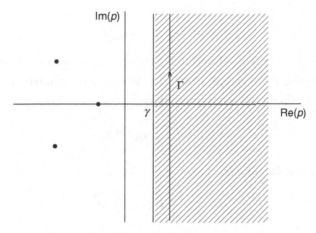

Figure 4.1. The complex p-plane. The hatched area shows the analytic region for $\bar{f}$ to the right of the line $\mathrm{Re}(p) = \gamma$. Solid circles represent singularities of $\bar{f}$.

complex p-plane on the right side of the vertical line, $\mathrm{Re}(p) = \gamma$. Thus, $\bar{f}(p)$ is analytic in the shaded area in Fig. 4.1.

4.1 INVERSION FORMULA

We may use what we learned from the complex Fourier transform to obtain the inversion formula for the Laplace transform. As $f = 0$ when $t < 0$, using the Heaviside step function $h(t)$, we define

$$g(t) = f(t)e^{-\gamma t}h(t), \qquad (4.3)$$

with γ sufficiently large to make g absolutely integrable. The Fourier integral theorem when applied to g gives

$$g(t) = \frac{1}{2\pi} \int_{-\infty}^{\infty} e^{-i\xi t} \int_{0}^{\infty} f(\tau)e^{-\gamma\tau+i\xi\tau} d\tau d\xi. \qquad (4.4)$$

Replacing g by $f(t)e^{-\gamma t}$,

$$f(t) = \frac{1}{2\pi} \int_{-\infty}^{\infty} e^{(\gamma t - i\xi)t} \int_{0}^{\infty} f(\tau)e^{-\gamma\tau+i\xi\tau} d\tau d\xi. \qquad (4.5)$$

Let

$$p = \gamma - i\xi, \quad dp = -i d\xi. \tag{4.6}$$

This change of variable alters our inversion integral from the line $-\infty < \xi < \infty$ to the line Γ going from $-i\infty$ to $i\infty$ in the analytic region in Fig. 4.1. Defining the inner integral as

$$\bar{f}(p) = \int_0^\infty f(t) e^{-pt} dt, \tag{4.7}$$

the inversion formula is

$$f(t) = \frac{1}{2\pi i} \int_\Gamma \bar{f}(p) e^{pt} dp. \tag{4.8}$$

The line Γ running vertically in the analytic region of $\bar{f}$ is called the Bromwich contour. From Eq. (4.1), when $\bar{f}$ is the Laplace transform of some function f, as $\mathrm{Re}(p) \to \infty, \bar{f} \to 0$. Now, given a function $\bar{f}$, we test if there is a semi-infinite region beyond $\mathrm{Re}(p) = \gamma$ where it is analytic. Then, we also test for the limit of $\bar{f}$ as $\mathrm{Re}(p) \to \infty$. If $\bar{f}$ passes both of these tests, we may attempt to invert it using the inversion integral. For the uniqueness of the inverse transform, we have to limit ourselves to functions, $f(t)$, that are piecewise continuous and discard "functions" that are undefined on certain finite intervals.

4.2 PROPERTIES OF THE LAPLACE TRANSFORM

First, let us consider a continuous function $f(t)$ for the purpose of illustrating the basic properties.

4.2.1 Linearity

From the definition given in Eq. (4.1), a linear combination of two functions has the transform

$$\mathcal{L}[f_1(t) + f_2(t)] = \bar{f}_1(p) + \bar{f}_2(p). \tag{4.9}$$

4.2.2 Scaling

Assuming a is positive, a change of scale of the time variable, t, to at results in

$$\mathcal{L}[f(at)] = \int_0^\infty f(at)e^{-pt}dt$$

$$= \frac{1}{a}\int_0^\infty f(t)e^{-pt/a}dt$$

$$= \frac{1}{a}\bar{f}(p/a). \tag{4.10}$$

4.2.3 Shifting

If the graph of the function is shifted by an amount a to the right,

$$\mathcal{L}[f(t-a)] = \int_a^\infty f(t-a)e^{-pt}dt$$

$$= \int_0^\infty f(t)e^{-p(t+a)}dt$$

$$= e^{-pa}\bar{f}(p). \tag{4.11}$$

The function $f(t)$ is zero when the argument is negative and $f(t-a)$ is zero when $t < a$.

4.2.4 Phase Factor

Here, "phase" is used in a generalized sense. What we have is the original $f(t)$ multiplied by the factor e^{-at}.

$$\mathcal{L}[f(t)e^{-at}] = \int_0^\infty f(t)e^{-(p+a)t}dt$$

$$= \bar{f}(p+a). \tag{4.12}$$

4.2.5 Derivative

When considering functions of time, it is customary to denote derivatives using "dots" over the variable.

$$\frac{df}{dt} = \dot{f}, \quad \frac{d^2f}{dt^2} = \ddot{f}, \quad \frac{d^n f}{dt^n} = f^{(n)}. \tag{4.13}$$

$$\mathcal{L}[\dot{f}(t)] = \int_0^\infty \dot{f}(t)e^{-pt}dt$$

$$= f(t)e^{-pt}\big|_0^\infty + p\int_0^\infty f(t)e^{-pt}dt$$

$$= p\bar{f}(p) - f(0). \tag{4.14}$$

Similarly,

$$\mathcal{L}[\ddot{f}(t)] = p\mathcal{L}[\dot{f}(t)] - \dot{f}(0)$$

$$= p^2\bar{f}(p) - pf(0) - \dot{f}(0). \tag{4.15}$$

$$\mathcal{L}[f^{(n)}(t)] = p^n\bar{f} - p^{n-1}f(0) - p^{n-2}\dot{f}(0) - \cdots - f^{n-1}(0). \tag{4.16}$$

These are important properties that allow us to convert differential equations into algebraic equations in the transformed form.

4.2.6 Integral

Consider the integral

$$F(t) = \int_0^t f(\tau)d\tau. \tag{4.17}$$

Using the relation for the derivative,

$$\mathcal{L}[f(t)] = p\mathcal{L}[F(t)] - F(0),$$

$$\mathcal{L}[F(t)] = \frac{1}{p}\bar{f}(p). \tag{4.18}$$

4.2.7 Power Factors

$$\mathcal{L}[t^n f(t)] = \int_0^\infty t^n f(t) e^{-pt} dt$$

$$= (-1)^n \frac{d^n}{dp^n} \int_0^\infty f(t) e^{-pt} dt$$

$$= (-1)^n \frac{d^n \bar{f}}{dp^n}. \tag{4.19}$$

4.3 TRANSFORMS OF ELEMENTARY FUNCTIONS

Recalling that all our functions are zero when $t < 0$,

$$\mathcal{L}[1] = \int_0^\infty e^{-pt} dt = \frac{1}{p}. \tag{4.20}$$

The integral of 1, namely, t has

$$\mathcal{L}[t] = -\frac{d}{dp}\frac{1}{p} = \frac{1}{p^2}. \tag{4.21}$$

Repeating the integration,

$$\mathcal{L}[t^n] = \frac{n!}{p^{n+1}}. \tag{4.22}$$

Using the result from the phase factor multiplication, we have

$$\mathcal{L}[e^{at}] = \frac{1}{p-a}, \quad \mathcal{L}[e^{iat}] = \frac{1}{p-ia}. \tag{4.23}$$

$$\mathcal{L}[\cos at] = \frac{1}{2}\left[\frac{1}{p-ia} + \frac{1}{p+ia}\right] = \frac{p}{p^2+a^2}, \quad \mathcal{L}[\sin at] = \frac{a}{p^2+a^2}. \tag{4.24}$$

Similarly,

$$\mathcal{L}[\cosh at] = \frac{1}{2}\left[\frac{1}{p-a} + \frac{1}{p+a}\right] = \frac{p}{p^2-a^2}, \quad \mathcal{L}[\sinh at] = \frac{a}{p^2-a^2}. \tag{4.25}$$

Recalling the definition of the Gamma function,

$$\Gamma(v+1) = \int_0^\infty t^v e^{-t} dt, \quad v > -1, \tag{4.26}$$

note that

$$\mathcal{L}[t^v] = \int_0^\infty t^v e^{-pt} dt,$$

$$= \frac{1}{p^{v+1}} \int_0^\infty \tau^v e^{-\tau} d\tau, \quad \tau = pt,$$

$$= \frac{\Gamma(v+1)}{p^{v+1}}. \tag{4.27}$$

When v is an integer, n, $\Gamma(n+1) = n!$, and we recover $\mathcal{L}[t^n] = n! p^{-(n+1)}$.

4.4 CONVOLUTION INTEGRAL

The convolution integral of two functions f and g under the Laplace transform is defined as

$$f * g(t) = \int_0^t f(t-\tau) g(\tau) d\tau. \tag{4.28}$$

A change of variable, $t - \tau = \tau'$, gives

$$f * g(t) = \int_0^t g(t-\tau') f(\tau') d\tau' = g * f(t), \tag{4.29}$$

which establishes the commutation property of the convolution integral.

Taking the Laplace transform

$$\mathcal{L}[f * g(t)] = \int_0^\infty e^{-pt} \int_0^t f(t-\tau) g(\tau) d\tau \, dt$$

$$= \int_0^\infty g(\tau) \int_\tau^\infty f(t-\tau) e^{-pt} dt \, d\tau, \quad t - \tau = \tau',$$

$$= \int_0^\infty g(\tau) \int_0^\infty f(\tau') e^{-p(\tau+\tau')} d\tau' \, d\tau$$

$$= \bar{f}(p) \bar{g}(p). \tag{4.30}$$

Thus, a given transform can be factored, and the factors after inversion can be convoluted to get the inverse of the original transform.

As an application of convolution, we evaluate the integral

$$I = \int_0^1 \frac{d\tau}{\tau^\alpha(1-\tau)^{1-\alpha}}.$$
(4.31)

First, we generalize the integral to obtain the convolution form

$$I(t) = \int_0^t \frac{d\tau}{\tau^\alpha(t-\tau)^{1-\alpha}}.$$
(4.32)

The Laplace transform of this integral as the products of the transforms

$$\mathcal{L}[t^{-\alpha}] = \Gamma(1-\alpha)/p^{1-\alpha},$$
(4.33)

$$\mathcal{L}[t^{\alpha-1}] = \Gamma(\alpha)/p^\alpha,$$
(4.34)

can be written as

$$\bar{I} = \Gamma(\alpha)\Gamma(1-\alpha)/p.$$
(4.35)

Inverting this,

$$I(t) = \Gamma(\alpha)\Gamma(1-\alpha).$$
(4.36)

Here, the right-hand side shows that the integral is independent of t, which can be verified by letting $\tau/t = \tau'$ in the preceding convolution integral.

4.5 INVERSION USING ELEMENTARY PROPERTIES

From a knowledge of the properties of the Laplace transform and the transforms of elementary functions, we may be able to invert many transforms. For example, to find

$$f(t) = \mathcal{L}^{-1}\left[\frac{e^{-2p}}{p^2+2p+10}\right],$$
(4.37)

we use the basic properties of the Laplace transform. The factor e^{-2p} represents a shift in time, and we can write

$$f(t) = g(t-2)h(t-2),$$
(4.38)

where

$$g(t) = \mathcal{L}^{-1}\left[\frac{1}{(p+1)^2+9}\right]. \qquad (4.39)$$

We may replace $(p+1)$ by p, by introducing a factor of e^{-t}. Thus,

$$g(t) = e^{-t}\mathcal{L}^{-1}\frac{1}{3}\left[\frac{3}{p^2+3^2}\right]$$

$$= \frac{1}{3}e^{-t}\sin 3t. \qquad (4.40)$$

Finally,

$$f(t) = \frac{1}{3}e^{(2-t)}\sin 3(t-2)h(t-2). \qquad (4.41)$$

Here, we have inserted the Heaviside step function h for clarity.

4.6 INVERSION USING THE RESIDUE THEOREM

When the given transform $\bar{f}(p)$ has only poles and no branch cuts, we use the inversion formula

$$f(t) = \frac{1}{2\pi i}\int_\Gamma \frac{e^{p(t-2)}}{p^2+2p+10}dp, \qquad (4.42)$$

where the vertical line Γ in the complex p-plane is selected to keep all the poles of the integrand on the left-hand side of it. The open contour Γ can be closed in two ways: When $(t-2)$ is positive, we extend it using a semi circle of infinite radius, C_∞, lying on the left-hand side of the plane as shown Fig. 4.2, and when $(t-2)$ is negative, we use a semicircle on the right-hand side of the plane. In the latter case, as there are no singularities inside the contour we obtain $f(t) = 0$ by the residue theorem. In the former case, the integral on this semicircle goes to zero by Jordan's lemma and by the residue theorem,

$$f(t) = \sum \text{Residues}, \qquad (4.43)$$

of the function $\bar{f}(p)e^{pt}$. Let us illustrate this using our previous example

$$\bar{f}e^{pt} = \frac{e^{(t-2)p}}{p^2+2p+10}. \qquad (4.44)$$

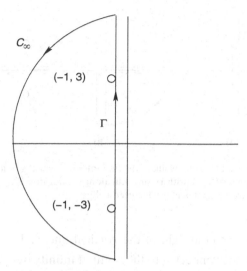

Figure 4.2. Closed contour to use with the residue theorem.

This function has poles at $(-1, \pm 3i)$ and

$$\text{Res}(-1+3i) = \frac{e^{(t-2)(-1+3i)}}{6i}, \quad \text{Res}(-1-3i) = \frac{e^{(t-2)(-1-3i)}}{-6i}. \quad (4.45)$$

The sum of the residues along with the fact that the integral is zero when $t < 2$, gives

$$f(t) = \frac{1}{3}e^{(2-t)}\sin 3(t-2)h(t-2). \quad (4.46)$$

4.7 INVERSION REQUIRING BRANCH CUTS

When the given transform requires branch cuts to create analytical domains, closing the Bromwich contour Γ using semicircles may not be feasible. Let us consider

$$f(t) = \mathcal{L}^{-1}[e^{-a\sqrt{P}}]. \quad (4.47)$$

The integrand in the inversion integral

$$f(t) = \frac{1}{2\pi i}\int_{\Gamma} e^{pt-a\sqrt{P}}\,dp \quad (4.48)$$

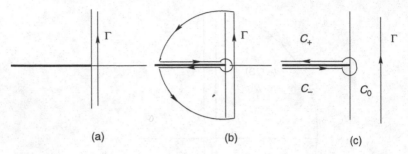

Figure 4.3. Closed contours to evaluate the line integral over Γ when branch cuts are present: (a) The original line Γ with a branch cut along the negative $\mathrm{Re}(p)$-axis, (b) closed contour using quarter circles, and (c) distorted Γ-line.

must be analytic to the right of the vertical line Γ. Then, the branch point $p = 0$ can be connected to the point at infinity by taking a branch cut, $-\infty < \mathrm{Re}(p) < 0$, which is shown as a thick line in Fig. 4.3(a). We may create a closed contour using two circular arcs, two lines (above and below the branch cut), and small circle around the origin as shown in Fig. 4.3(b). As there are no singularities inside this closed contour, we may also distort Γ by bending it around the branch point $p = 0$ to get an equivalent contour as shown in Fig. 4.3(c).

We will use the contour shown in Fig. 4.3(c) for our calculation. We have

$$f(t) = \frac{1}{2\pi i} \int_{\Gamma} e^{pt - a\sqrt{p}} dp$$

$$= \frac{1}{2\pi i} \left[\int_{C_-} + \int_{C_0} + \int_{C_+} \right] e^{pt - a\sqrt{p}} dp. \qquad (4.49)$$

On C_0,

$$p = \epsilon e^{i\theta}, \quad dp = \epsilon e^{i\theta} i d\theta, \qquad (4.50)$$

and

$$\int_{C_0} = \int_{-\pi}^{\pi} \exp[\epsilon e^{i\theta} - \sqrt{\epsilon} e^{i\theta/2}] \epsilon e^{i\theta} i d\theta. \qquad (4.51)$$

As $\epsilon \to 0$, this integral goes to zero.

On C_+,

$$p = re^{i\pi}, \quad \sqrt{p} = \sqrt{r}e^{i\pi/2} = i\sqrt{r}, \quad dp = -dr, \qquad (4.52)$$

and

$$\int_{C_+} = -\int_0^\infty e^{-rt - ia\sqrt{r}} dr. \qquad (4.53)$$

On C_-,

$$p = re^{-i\pi}, \quad \sqrt{p} = \sqrt{r}e^{-i\pi/2} = -i\sqrt{r}, \quad dp = -dr, \qquad (4.54)$$

and

$$\int_{C_-} = -\int_\infty^0 e^{-rt + ia\sqrt{r}} dr. \qquad (4.55)$$

Thus, combining the line integrals, we obtain

$$f(t) = \frac{1}{\pi} \int_0^\infty e^{-rt} \sin a\sqrt{r} \, dr. \qquad (4.56)$$

Next, we use a change of variable

$$x^2 = r, \quad dr = 2x dx, \qquad (4.57)$$

to get

$$f(t) = \frac{2}{\pi} \int_0^\infty e^{-x^2 t} \sin(ax) x \, dx. \qquad (4.58)$$

This may be viewed as the Fourier Sine transform with respect to the transform variable a. We have

$$\mathcal{F}_s[xe^{-b^2 x^2}, x \to a] = \frac{1}{(\sqrt{2}b)^3} e^{-a^2/(4b^2)}. \qquad (4.59)$$

Finally,

$$\mathcal{L}^{-1}[e^{-a\sqrt{p}}] = \frac{a}{\sqrt{4\pi t^3}} e^{-a^2/(4t)}. \qquad (4.60)$$

Integrating both sides with respect to a, from a to ∞,

$$\mathcal{L}^{-1}\left[\frac{e^{-a\sqrt{p}}}{\sqrt{p}}\right] = \frac{1}{\sqrt{\pi t}} e^{-a^2/(4t)}. \qquad (4.61)$$

One more integration with respect to a gives

$$\mathcal{L}^{-1}\left[\frac{e^{-a\sqrt{p}}}{p}\right] = \frac{1}{\sqrt{\pi t}}\int_a^\infty e^{-a^2/(4t)}\,da$$

$$= \frac{1}{\sqrt{\pi t}}\int_{a/(2\sqrt{t})}^\infty e^{-\tau^2}2\sqrt{t}\,d\tau, \quad \tau = \frac{a}{2\sqrt{t}},$$

$$= \frac{2}{\sqrt{\pi}}\int_{a/(2\sqrt{t})}^\infty e^{-\tau^2}\,d\tau$$

$$= \operatorname{erfc}\left[\frac{a}{2\sqrt{t}}\right]. \tag{4.62}$$

As we will see, these Laplace transforms with branch cuts appear frequently in heat conduction problems.

4.8 THEOREMS OF TAUBER

From the behavior of $\bar{f}(p)$ for large values of $|p|$ and small values of $|p|$, we may infer the behavior of $f(t)$ for small values of t and large values of t without actually inverting the transform. These results are known as Tauber's theorems (or Tauberian theorems).

4.8.1 Behavior of $f(t)$ as $t \to 0$

Assuming $f(t)$ is bounded by $e^{\gamma t}$ and it has the form

$$f(t) = c_0 + c_1 t + c_2 t^2 + \cdots, \tag{4.63}$$

near $t = 0$, we have

$$\bar{f}(p) = \frac{c_0}{p} + \frac{c_1}{p^2} + \frac{2c_2}{p^3} + \cdots. \tag{4.64}$$

From this

$$\lim_{p\to\infty} p\bar{f}(p) = c_0 = \lim_{t\to 0} f(t). \tag{4.65}$$

4.8.2 Behavior of $f(t)$ as $t \to \infty$

Assuming $f(t)$ is bounded as $t \to \infty$, we have

$$p\bar{f}(p) - f(\infty) = \int_0^\infty [f(t) - f(\infty)]e^{-pt}pdt. \qquad (4.66)$$

Substituting $x = pt$,

$$p\bar{f}(p) - f(\infty) = \int_0^\infty [f(x/p) - f(\infty)]e^{-x}dx. \qquad (4.67)$$

$$\lim_{p \to 0} p\bar{f}(p) = f(\infty) = \lim_{t \to \infty} f(t). \qquad (4.68)$$

In many engineering applications, we are interested in the short-time or long-time response of a system; the Tauberian theorems lead to these results without a detailed calculation of the inverse for all times.

4.9 APPLICATIONS OF LAPLACE TRANSFORM

Some typical applications of the Laplace transform are described in the following sections.

4.9.1 Ordinary Differential Equations

Consider the differential equation

$$u^{(n)} + a_1 u^{(n-1)} + \cdots + a_{n-1}u^{(1)} + a_n u = f(t), \qquad (4.69)$$

where a_j are constants and

$$u^{(j)} = \frac{d^j u}{dt^j}, \quad j = 1, 2, \ldots, n, \qquad (4.70)$$

with the homogeneous initial conditions

$$u(0) = u^{(1)}(0) = \cdots = u^{(n-1)}(0) = 0. \qquad (4.71)$$

Taking the Laplace transform of both sides, we get

$$[p^n + a_1 p^{n-1} + \cdots + a_n]\bar{u} = \bar{f}. \qquad (4.72)$$

The Laplace transform of the solution is

$$\bar{u} = \bar{g}\bar{f}, \tag{4.73}$$

where

$$\bar{g} = \frac{1}{p^n + a_1 p^{n-1} + \cdots + a_n}. \tag{4.74}$$

The nth degree polynomial in the denominator, known as the characteristic polynomial, may be factored in the form

$$p^n + a_1 p^{n-1} + \cdots + a_n = (p - p_1)^{j_1} (p - p_2)^{j_2} \cdots (p - p_m)^{j_m}, \tag{4.75}$$

where

$$j_1 + j_2 + \cdots + j_m = n. \tag{4.76}$$

Here, p_1, etc. are the zeros of the characteristic polynomial. These may be complex numbers. We also include the possibility of repeated zeros. A partial fraction expansion of $\bar{g}$ can be expressed as

$$\bar{g} = \frac{a^{(1)} + a^{(2)}(p - p_1) + \cdots + a^{(j_1)}(p - p_1)^{j_1 - 1}}{(p - p_1)^{j_1}} + \cdots. \tag{4.77}$$

Using the elementary result

$$\mathcal{L}^{-1} \frac{1}{p - p_1} = e^{p_1 t}, \tag{4.78}$$

$$g(t) = \mathcal{L}^{-1}[\bar{g}] = \left[\frac{a^{(1)} t^{j_1}}{j_1!} + \frac{a^{(2)} t^{j_1 - 1}}{(j_1 - 1)!} + \cdots + a^{(j_1)} \right] e^{p_1 t} + \cdots. \tag{4.79}$$

Once $g(t)$ is known, the solution can be expressed in the convolution form

$$u(t) = \int_0^t g(t - \tau) f(\tau) d\tau, \tag{4.80}$$

where $g(t - \tau)$ can be recognized as the Green's function for our differential equation with homogeneous initial conditions.

When the coefficients a_j are real, the zeros of the characteristic polynomial come in pairs of complex conjugate numbers, in the form

$$p_1 = \lambda_1 + i\omega_1, \quad p_2 = \lambda_1 - i\omega_1, \quad \text{etc.} \tag{4.81}$$

In this case, terms of the form

$$\frac{a^{(1)}}{p - p_1} + \frac{a^{(2)}}{p - p_2}, \tag{4.82}$$

can be combined to obtain

$$\frac{b^{(1)}(p - \lambda_1) + b^{(2)}\omega_1}{(p - \lambda_1)^2 + \omega_1^2}, \tag{4.83}$$

which can be inverted in terms $e^{\lambda_1 t} \cos \omega_1 t$ and $e^{\lambda_1 t} \sin \omega_1 t$.

Example: Vibrations

For a spring and mass system with mass m and spring constant k, the equation of motion may be written as

$$m\ddot{u} + ku = f(t), \quad u(0) = \dot{u}(0) = 0, \tag{4.84}$$

where f is the force exciting the system. Using

$$\omega^2 = k/m, \tag{4.85}$$

the Laplace transform gives

$$\bar{u} = \bar{g}\bar{f}, \tag{4.86}$$

where

$$\bar{g} = \frac{1}{m} \frac{1}{p^2 + \omega^2}. \tag{4.87}$$

Inverting,

$$g(t) = \frac{1}{m\omega} \sin \omega t, \tag{4.88}$$

and the solution is

$$u(t) = \frac{1}{m\omega} \int_0^t \sin \omega (t - \tau) f(\tau) d\tau. \tag{4.89}$$

Example: Higher-Order Differential Equation

Consider the differential equation

$$\frac{d^4u}{dt^4} - 6\frac{d^3u}{dt^3} + 14\frac{d^2u}{dt^2} - 16\frac{du}{dt} + 8u = t, \qquad (4.90)$$

with the initial conditions

$$\frac{d^3u}{dt^3}(0) = \frac{d^2u}{dt^2}(0) = \frac{du}{dt}(0) = u(0) = 0. \qquad (4.91)$$

Taking the Laplace transform

$$\bar{u} = \frac{1}{p^2(p^4 - 6p^3 + 14p^2 - 16p + 8)}. \qquad (4.92)$$

By inspection, we find $p = 2$ is a double zero. Removing the factors $(p-2)^2$ and p^2, what remains is $p^2 - 2p + 2$, which can be written as $[(p-1)^2 + 1]$. Thus,

$$\bar{u} = \frac{1}{p^2(p-2)^2[(p-1)^2 + 1]}. \qquad (4.93)$$

A partial fraction expansion of the right-hand side expression is

$$\frac{1}{p^2(p-2)^2[(p-1)^2 + 1]} = \frac{A_1}{p} + \frac{A_2}{p^2} + \frac{B_1}{(p-2)}$$

$$+ \frac{B_2}{(p-2)^2} + \frac{C_1 p + C_2}{[(p-1)^2 + 1]}, \qquad (4.94)$$

where the constants $A_1, \ldots, C_2$ have to be found. Multiplying both sides by p^2 and setting $p = 0$, we find

$$A_2 = \frac{1}{8}, \qquad (4.95)$$

and multiplying with $(p-2)^2$ and setting $p = 2$,

$$B_2 = \frac{1}{8}. \qquad (4.96)$$

The quadratic expression $[(p-1)^2 + 1]$ can be factored as $(p-1-i)$ $(p-1+i)$. Multiplying by the quadratic and setting $p = 1 \pm i$ gives

$$C_1 = 0, \quad C_2 = \frac{1}{4}. \qquad (4.97)$$

The remaining constants are found as

$$A_1 = \frac{1}{4}, \quad B_1 = -\frac{1}{4}, \tag{4.98}$$

by setting p to two arbitrary numbers other than the zeros of the denominator of $\bar{u}$, say, $p = 1$ and $p = -1$. Finally,

$$\bar{u} = \frac{1}{8}\left[\frac{2}{p} + \frac{1}{p^2} - \frac{2}{p-2} + \frac{1}{(p-2)^2} + \frac{2}{(p-1)^2+1}\right], \tag{4.99}$$

which has the inverse

$$u = \frac{1}{8}[2 + t - 2e^{2t} + te^{2t} + 2e^t \sin t]. \tag{4.100}$$

4.9.2 Boundary Value Problems

In some problems, the differential equation is supplemented with boundary conditions, instead of the initial conditions we have assumed so far. In such cases, we substitute unknown constants for the missing initial conditions while taking the Laplace transform, and solve for these constants using the unused boundary conditions after inverting the solution. Most of the time, differential equations containing an unknown space-dependent variable come with boundary conditions.

4.9.3 Partial Differential Equations

We consider two examples of partial differential equations: transient heat conduction in a semi-infinite bar and wave propagation in a semi-infinite string.

Example: Transient Heat Conduction

With u representing the relative temperature above the ambient, the heat conduction in a semi-infinite bar is described by the equation

$$\kappa\frac{\partial^2 u}{\partial x^2} = \frac{\partial u}{\partial t}, \quad 0 < x < \infty, \quad 0 < t < \infty, \tag{4.101}$$

where κ is the diffusivity. We assume

$$u(x,0) = 0, \quad u(0,t) = f(t), \quad u(x,t) \to 0, \quad \text{as} \quad x \to \infty. \tag{4.102}$$

Taking the transform

$$\mathcal{L}[u(x,t), t \to p] = \bar{u}(x,p), \tag{4.103}$$

$$\kappa \frac{\partial^2 \bar{u}}{\partial x^2} = p\bar{u}. \tag{4.104}$$

This second-order ordinary differential equation in x has the general solution

$$\bar{u} = A e^{-\sqrt{p/\kappa}x} + B e^{\sqrt{p/\kappa}x}. \tag{4.105}$$

As $u \to 0$ as $x \to \infty$, $B = 0$.

The end condition $u(0,t) = f(t)$ transforms into $\bar{u}(0,p) = \bar{f}$, and we find

$$A = \bar{f}. \tag{4.106}$$

Then

$$\bar{u}(x,p) = \bar{f} e^{-\sqrt{p/\kappa}x}. \tag{4.107}$$

Using convolution, the solution has the form

$$u(x,t) = \int_0^t g(t-\tau)f(\tau)d\tau, \tag{4.108}$$

where the Green's function g is given by

$$g(t) = \mathcal{L}^{-1}[e^{-\sqrt{p/\kappa}x}]$$
$$= \frac{1}{\sqrt{4\pi\kappa t^3}} e^{-x^2/(4\kappa t)}, \tag{4.109}$$

where we have used Eq. (4.60). In the case of a sudden increase in temperature, T_0, at the end of the bar $x = 0$, we have

$$f(t) = T_0, \tag{4.110}$$

$$\bar{u}(x,p) = \frac{T_0}{p} e^{-\sqrt{p/\kappa}x}, \tag{4.111}$$

and we may invert this directly using Eq. (4.62), without resorting to convolution, in the form

$$u(x,t) = T_0 \operatorname{erfc}\left[\frac{x}{\sqrt{4\kappa t}}\right]. \tag{4.112}$$

Example: Wave Propagation in a Semi-infinite String

We consider a taut string occupying $0 < x < \infty$. The differential equation describing the perpendicular displacement is

$$\frac{\partial^2 u}{\partial t^2} = c^2 \frac{\partial^2 u}{\partial x^2}, \tag{4.113}$$

where c is the wave speed that depends on the tension in the string and the density of the string material. We assume the boundary and initial conditions:

$$u(x,0) = \dot{u}(x,0) = 0, \quad u(0,t) = f(t), \quad u(x,t) \to 0 \quad \text{as} \quad x \to \infty. \tag{4.114}$$

Taking the transform, we obtain the second-order differential equation in x,

$$\frac{\partial^2 \bar{u}}{\partial x^2} = \frac{p^2}{c^2}\bar{u}, \tag{4.115}$$

which has the solution

$$\bar{u} = Ae^{-px/c} + Be^{px/c}. \tag{4.116}$$

As it is required for the solution to vanish at infinity, $B = 0$. Using the end condition, $\bar{u}(0,p) = \bar{f}$, we find

$$A = \bar{f}, \tag{4.117}$$

$$\bar{u} = \bar{f}e^{-px/c}. \tag{4.118}$$

Noting that the effect of the exponential factor is a shift in time,

$$u(x,t) = f(t - x/c)h(t - x/c), \tag{4.119}$$

where we have used the Heaviside step function for clarity. It can be seen that the disturbance imparted to the end, $x = 0$, travels along the string with a speed of c.

Example: D'Alembert Solution for Waves in an Infinite String

In an infinite string, if the displacement and velocities are prescribed initially, we have

$$\frac{\partial^2 u}{\partial t^2} = c^2 \frac{\partial^2 u}{\partial x^2}, \tag{4.120}$$

$$u(x,0) = f(x), \quad \dot{u}(x,0) = g(x). \tag{4.121}$$

Taking the transform

$$p^2 \bar{u} - pf(x) - g(x) = c^2 \frac{\partial^2 \bar{u}}{\partial x^2}, \tag{4.122}$$

where, as before,

$$\bar{u} = \mathcal{L}[u(x,t), t \to p]. \tag{4.123}$$

Next, we take the Fourier transform of $\bar{u}(x,p)$,

$$\bar{U}(\xi,p) = \mathcal{F}[\bar{u}(x,p), x \to \xi]. \tag{4.124}$$

Now the differential equation is

$$(p^2 + c^2\xi^2)\bar{U} = pF(\xi) + G(\xi),$$
$$\bar{U} = \frac{pF(\xi) + G(\xi)}{p^2 + c^2\xi^2}. \tag{4.125}$$

Inverting the Laplace transform first and then the Fourier transform,

$$U = \frac{1}{2}\left[e^{ict\xi} + e^{-ict\xi}\right]F(\xi) + \frac{1}{2ic\xi}\left[e^{ict\xi} - e^{-ict\xi}\right]G(\xi)$$
$$= \frac{1}{2}\left[e^{ict\xi} + e^{-ict\xi}\right]F(\xi) + \frac{1}{2}\int_{-t}^{t} e^{ic\tau\xi}G(\xi)d\tau,$$

$$u(x,t) = \frac{1}{2}\left[f(x+ct) + f(x-ct) + \int_{-t}^{t} g(x-c\tau)d\tau \right]$$

$$= \frac{1}{2}\left[f(x+ct) + f(x-ct) + \frac{1}{c}\int_{x-ct}^{x+ct} g(x')dx' \right], \qquad (4.126)$$

where we have used $x' = x - c\tau$.

The traditional derivation of this result uses new variables (not to be confused with the Fourier variable),

$$\xi = x + ct, \quad \eta = x - ct, \qquad (4.127)$$

to have

$$\frac{\partial}{\partial x} = \frac{\partial}{\partial \xi} + \frac{\partial}{\partial \eta}, \quad \frac{1}{c}\frac{\partial}{\partial t} = \frac{\partial}{\partial \xi} - \frac{\partial}{\partial \eta}. \qquad (4.128)$$

In these new variables, the wave equation becomes

$$\frac{\partial^2 u}{\partial \xi \partial \eta} = 0. \qquad (4.129)$$

This equation, after integration, gives the solution in terms of two arbitrary functions, F and G, in the form,

$$u = F(\xi) + G(\eta) = F(x+ct) + G(x-ct). \qquad (4.130)$$

From the initial conditions,

$$F(x) + G(x) = f(x), \quad F'(x) - G'(x) = g(x)/c. \qquad (4.131)$$

Integrating the second relation,

$$F(x) - G(x) = \frac{1}{c}\int_{0}^{x} g(x')dx', \qquad (4.132)$$

and solving for F and G,

$$F(x) = \frac{1}{2}\left[f(x) + \frac{1}{c}\int_{0}^{x} g(x')dx' \right], \qquad (4.133)$$

$$G(x) = \frac{1}{2}\left[f(x) - \frac{1}{c}\int_{0}^{x} g(x')dx' \right]. \qquad (4.134)$$

With these

$$u = \frac{1}{2}\left[f(x+ct)+f(x-ct)+\frac{1}{c}\int_{x-ct}^{x+ct} g(x')dx'\right]. \qquad (4.135)$$

The new variables, $\xi = x+ct$ and $\eta = x-ct$, are called the characteristic variables. In the (x,t)-plane, a fixed value of ξ or η corresponds to a straight line known as the characteristic line. The initial conditions propagate along the characteristic lines.

4.9.4 Integral Equations

Integral equations containing convolution integrals may be solved using the Laplace transform. Of course, the convolution integrals must be suitable for the Laplace transform.

Example: Volterra Equation

Consider the integral equation

$$u(t) + \lambda \int_0^t g(t-\tau)u(\tau)d\tau = f(t), \qquad (4.136)$$

where λ is a constant. Kernels of the form $g(t-\tau)$ are called hereditary kernels as they express the history of the unknown, u. Taking the transform

$$(1+\lambda\bar{g})\bar{u} = \bar{f}, \qquad (4.137)$$

$$\bar{u} = \frac{\bar{f}}{1+\lambda\bar{g}}. \qquad (4.138)$$

If we can invert this expression, we obtain the solution. In the special case,

$$g(t) = e^{-t}, \quad \bar{g} = \frac{1}{p+1}, \qquad (4.139)$$

$$\bar{u} = \frac{(p+1)\bar{f}}{p+\lambda+1} = \bar{f} - \lambda\frac{\bar{f}}{p+\lambda+1}, \qquad (4.140)$$

which can be inverted to obtain

$$u(t) = f(t) - \lambda \int_0^t e^{-(\lambda+1)(t-\tau)} f(\tau) d\tau. \qquad (4.141)$$

Example: Abel Equation

The Abel equation,

$$\int_0^t \frac{u(\tau) d\tau}{\sqrt{t-\tau}} = f(t), \qquad (4.142)$$

transforms to

$$\frac{\sqrt{\pi}}{\sqrt{p}} \bar{u} = \bar{f}, \qquad (4.143)$$

where we have used

$$\mathcal{L}\left[\frac{1}{\sqrt{t}}\right] = \sqrt{\frac{\pi}{p}}, \qquad (4.144)$$

$$\bar{u} = \frac{p\bar{f}}{\sqrt{\pi p}}. \qquad (4.145)$$

Inversion yields

$$u(t) = \frac{1}{\pi} \frac{d}{dt} \int_0^t \frac{f(\tau) d\tau}{\sqrt{t-\tau}}. \qquad (4.146)$$

Example: Generalized Abel Equation

A slight modification of the Abel equation has the form

$$\int_0^t \frac{u(\tau) d\tau}{\sqrt{t^2 - \tau^2}} = f(t). \qquad (4.147)$$

We may introduce new variables

$$x = t^2, \quad \xi = \tau^2, \quad d\tau = d\xi/(2\sqrt{\xi}), \qquad (4.148)$$

to obtain

$$\int_0^x \frac{u(\sqrt{\xi}) d\xi}{2\sqrt{\xi}\sqrt{x-\xi}} = f(\sqrt{x}). \qquad (4.149)$$

Let

$$v(\xi) = \frac{u(\sqrt{\xi})}{2\sqrt{\xi}}, \quad g(x) = f(\sqrt{x}). \qquad (4.150)$$

Now we have the Abel equation,

$$\int_0^t \frac{v(\xi)d\xi}{\sqrt{x-\xi}} = g(x), \tag{4.151}$$

which has the solution

$$v(x) = \frac{1}{\pi}\frac{d}{dx}\int_0^x \frac{g(\xi)d\xi}{\sqrt{x-\xi}}. \tag{4.152}$$

In terms of the original variables,

$$u(t) = \frac{2}{\pi}\frac{d}{dt}\int_0^t \frac{f(\tau)\tau d\tau}{\sqrt{t^2-\tau^2}}. \tag{4.153}$$

4.9.5 Cagniard–De Hoop Method

To solve partial differential equations involving infinite spaces and time, we may use the Laplace and Fourier transforms simultaneously as we have done with the multiple Fourier transforms. Once the transform of the unknown function is isolated, it can be inverted using the inverse integrals sequentially. The Cagniard-De Hoop method uses the Fourier inverse first, and the line integral from $-\infty$ to ∞ is distorted to look like a Laplace transform. In this way the inverse of the two transforms are obtained without any contour integration if there are no singularities between the original contour and the distorted contour.

We illustrate this method using a semi-infinite medium, $-\infty < x < \infty, 0 < y < \infty$ at a quiescent pressure, p_0. At time, $t = 0$, an additional pressure of magnitude $P_0 f(x)$ is applied on the surface $y = 0$. It is assumed that the pressure fluctuations $p(x,y,t)$ above the ambient pressure p_0 satisfy

$$c^2\nabla^2 p = \frac{\partial^2 p}{\partial t^2}, \tag{4.154}$$

where c is the speed of sound. The boundary conditions are

$$p(x,0,t) = P_0 f(x)h(t), \quad p \to 0 \quad \text{as} \quad r \to \infty, \tag{4.155}$$

where $r = \sqrt{x^2 + y^2}$.

To avoid confusion, we use s for the Laplace variable, instead of p, and let

$$\bar{p}(x,y,s) = \mathcal{L}[p(x,y,t), t \to s].\qquad(4.156)$$

After taking the Laplace transform, we have

$$(\nabla^2 - k^2 s^2)\bar{p} = 0, \quad \bar{p}(x,0,s) = \frac{P_0 f(x)}{s},\qquad(4.157)$$

where $k = 1/c$. Next, we take the Fourier transform of the governing equation and the boundary condition to get

$$\frac{\partial^2 \bar{P}}{\partial y^2} - (\xi^2 + k^2 s^2)\bar{P} = 0, \quad \bar{P}(\xi,0,s) = \frac{P_0 F(\xi)}{s},\qquad(4.158)$$

where

$$\bar{P} = \mathcal{F}[\bar{p}, x \to \xi], \quad F = \mathcal{F}[f, x \to \xi].\qquad(4.159)$$

Solution of this equation vanishing at infinity is

$$\bar{P} = Ae^{-\sqrt{\xi^2 + k^2 s^2}\, y},\qquad(4.160)$$

where the square root has to be selected to have the real part of $\sqrt{\xi^2 + k^2 s^2}$ positive. This can be accomplished by introducing branch cuts from iks to $i\infty$ and from $-iks$ to $-i\infty$ in the complex ξ-plane.

Using the boundary condition, we find

$$A = P_0 F(\xi)/s.\qquad(4.161)$$

Thus,

$$\bar{P}(\xi,y,s) = \frac{P_0 F(\xi)}{s} e^{-\sqrt{\xi^2 + k^2 s^2}\, y}.\qquad(4.162)$$

Recalling the Green's functions, we express the solution in the convolution form,

$$p(x,y,t) = P_0 \int_{-\infty}^{\infty} g(x - x', y, t) f(x')dx',\qquad(4.163)$$

where

$$\bar{G} = \mathcal{F}\mathcal{L}[g(x,y,t)] = \mathcal{F}[\bar{g}(x,y,s)] = \frac{1}{\sqrt{2\pi}s} e^{-\sqrt{\xi^2 + k^2 s^2}\, y}.\qquad(4.164)$$

where we have inserted $1/\sqrt{2\pi}$ to avoid this factor in the anticipated convolution integral for Fourier transforms.

A direct approach for inverting $\bar{G}$ requires a Fourier inversion integral and a Laplace inversion integral. Using the Cagniard-De Hoop method, we can avoid actual integrations. The Fourier inverse is written as

$$\bar{g}(x,y,s) = \frac{1}{2\pi} \int_{-\infty}^{\infty} e^{-\sqrt{\xi^2 + k^2 s^2}\, y - i\xi x}\, \frac{d\xi}{s}. \qquad (4.165)$$

A change of variable, $\xi \to s\xi$, gives

$$\bar{g}(x,y,s) = \frac{1}{2\pi} \int_{-\infty}^{\infty} e^{-s(\sqrt{\xi^2 + k^2}\, y - i\xi x)}\, d\xi. \qquad (4.166)$$

In Fig. 4.4, this integration is over the line, C. In the new ξ-plane the branch points are at $\pm ik$. We introduce new parameter t (which will be later identified as time) to distort this line to Γ, by defining

$$\sqrt{\xi^2 + k^2}\, y - i\xi x = t,$$
$$(\xi^2 + k^2)y^2 = (t + i\xi x)^2,$$
$$\xi^2 + \frac{2itx}{r^2}\xi + \frac{k^2 y^2 - t^2}{r^2} = 0, \quad r^2 = x^2 + y^2. \qquad (4.167)$$

Solving for ξ, we get

$$\xi = \pm \frac{\sqrt{t^2 - k^2 r^2}}{r} \sin\theta - i\frac{t}{r}\cos\theta, \qquad (4.168)$$

where

$$\sin\theta = y/r, \quad \cos\theta = x/r. \qquad (4.169)$$

Let us distinguish the two roots as

$$\xi_+ = +\frac{\sqrt{t^2 - k^2 r^2}}{r} \sin\theta - i\frac{t}{r}\cos\theta, \qquad (4.170)$$

$$\xi_- = -\frac{\sqrt{t^2 - k^2 r^2}}{r} \sin\theta - i\frac{t}{r}\cos\theta. \qquad (4.171)$$

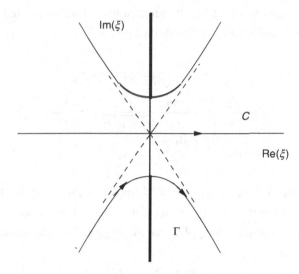

Figure 4.4. Distortion of the contour C to Γ in the Cagniard-De Hoop method.

Note that as t varies from kr to ∞, the real part of ξ_+ varies from 0 to ∞, and the imaginary part varies from $-k\cos\theta$ to $-\infty$. This is represented by the right half of the lower hyperbola in Fig. 4.4. For large t, the slope of the hyperbola asymptotically approaches $-\cot\theta$. The values of ξ from $-\infty$ to 0 is covered by t going from ∞ to kr and the asymptote for the left half of the lower hyperbola is $\cot\theta$.

The integral representation, Eq. (4.166), in terms of t is

$$\bar{g}(x,y,s) = \frac{1}{2\pi}\left[\int_\infty^{cr} e^{-st}\frac{d\xi_-}{dt}dt + \int_{cr}^\infty e^{-st}\frac{d\xi_+}{dt}dt\right]$$

$$= \frac{1}{2\pi}\int_{cr}^\infty\left[\frac{d\xi_+}{dt} - \frac{d\xi_-}{dt}\right]e^{-st}dt. \qquad (4.172)$$

We recognize this expression as a Laplace transform with respect to time, and the function $g(x,y,t)$, which is the function we are looking for, is the factor of e^{-st} in the integrand. Then,

$$g(x,y,t) = \frac{1}{2\pi}\left[\frac{d\xi_+}{dt} - \frac{d\xi_-}{dt}\right]h(t-kr), \qquad (4.173)$$

where the Heaviside step function indicates a shift as our integrals start from kr. Evaluation of the derivatives yields

$$\frac{d\xi_+}{dt} = \frac{\sin\theta}{r}\frac{t}{\sqrt{t^2 - k^2 r^2}} - i\frac{\cos\theta}{r}, \tag{4.174}$$

$$\frac{d\xi_+}{dt} = -\frac{\sin\theta}{r}\frac{t}{\sqrt{t^2 - k^2 r^2}} - i\frac{\cos\theta}{r}, \tag{4.175}$$

$$g(x,y,t) = \frac{1}{\pi}\frac{\sin\theta}{r}\frac{t}{\sqrt{t^2 - k^2 r^2}}. \tag{4.176}$$

The Green's function we have found has the property that it is the solution when $P_0 f(x) = P_0 \delta(x)$, a singular compressive force of magnitude P_0 suddenly applied at the origin. The pressure for this case is given by

$$p = \frac{P_0}{\pi}\frac{\sin\theta}{r}\frac{t}{\sqrt{t^2 - k^2 r^2}}. \tag{4.177}$$

When $t \to \infty$, we reach a steady-state solution,

$$p = \frac{P_0}{\pi}\frac{\sin\theta}{r}, \tag{4.178}$$

which is

$$p = -\frac{P_0}{\pi}\,\text{Im}(1/z), \quad z = x + iy, \tag{4.179}$$

and it satisfies the Laplace equation. In our original wave equation, the right-hand term vanishes for steady-state solutions. For a concentrated applied force of P_0, we should have

$$P_0 = \lim_{r \to 0}\int_0^{\pi/2} 2pr\sin\theta. \tag{4.180}$$

As we can verify, our solution satisfies this equilibrium condition.

The Cagniard–De Hoop method has found numerous applications in wave propagation problems in elastic media. The books by Fung (1965) and Ewing, Jardetzky and Press (1957) provide further citations and solutions.

4.10 SEQUENCES AND THE Z-TRANSFORM

Consider an infinite sequence of numbers

$$\{a_n\} = \{a_0, a_1, \ldots, a_n, \ldots\}. \tag{4.181}$$

We construct the so-called Z-transform of this sequence by multiplying a_n by z^n and adding. Here, z is a complex number, and it is assumed that the sequence is such that the sum converges for a suitable radius $|z|$.

$$\mathcal{Z}\{a_n\} = Z = \sum_0^\infty a_n z^n. \tag{4.182}$$

For example, the sequence

$$\{1\} = \{1, 1, \ldots, 1, \ldots\} \tag{4.183}$$

gives

$$Z = 1 + z + z^2 + \cdots = \frac{1}{1-z}, \tag{4.184}$$

provided $|z| < 1$. The sequence

$$\{a^n\} = \{1, a, a^n, \ldots\} \tag{4.185}$$

has

$$Z = 1 + az + a^2 z^2 + \cdots = \frac{1}{1-az}. \tag{4.186}$$

Our definition is slightly different from the commonly used definition – we use z instead of z^{-1} in our expansions.

The connection between the Z-transform and the Laplace transform will become clear if we convert the sequence into a discontinuous function of time. Between $t = 0$ and $t = 1$, the function has the value of a_0; between $t = 1$ and $t = 2$, it has the value of a_1, and so on. This is shown in Fig. 4.5. Thus, the discontinuous function $f(t)$ can be written as

$$f(t) = a_0[h(t) - h(t-1)] + a_1[h(t-1) - h(t-2)] + \cdots, \tag{4.187}$$

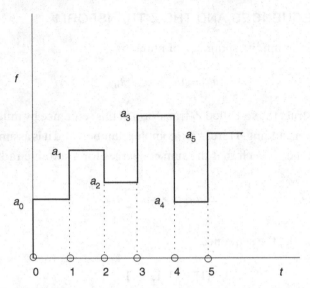

Figure 4.5. Representation of a sequence as a discontinuous function.

where h is the Heaviside step function. The Laplace transform,

$$\mathcal{L}[h(t-n)-h(t-n-1)] = \int_{n}^{n+1} e^{-pt}dt$$

$$= \frac{1}{p}\left[e^{-np} - e^{-(n+1)p}\right] = \frac{e^{-np}}{p}(1-e^{-p}).$$

(4.188)

Using this, we obtain the Laplace transform of our discontinuous function,

$$\bar{f} = \frac{1}{p}(1-e^{-p})\sum_{0}^{\infty} a_n e^{-np}.$$

(4.189)

A change of variable,

$$z = e^{-p},$$

(4.190)

makes

$$\bar{f} = \frac{z-1}{\text{Log } z}Z,$$

(4.191)

where

$$Z = \sum_{0}^{\infty} a_n z^n \qquad (4.192)$$

is the Z-transform. In choosing $\text{Log } z$, we have selected a branch along the negative axis of $\text{Re } (z)$.

Our elementary examples show that

$$\mathcal{Z}^{-1}\left[\frac{1}{1-az}\right] = \{a^n\}. \qquad (4.193)$$

In a more general case, the inverse Z-transform of a function $F(z)$, which is analytic within a circle around $z = 0$, is the coefficient of z^n in the Taylor series expansion of $F(z)$.

By differentiating Eq. (4.193) with respect to a, we find

$$\mathcal{Z}^{-1}\left[\frac{z}{(1-az)^2}\right] = \{na^{n-1}\}. \qquad (4.194)$$

Similarly, by differentiating Eq. (4.186) with respect to z and identifying the coefficient of z^n, we get

$$\mathcal{Z}^{-1}\left[\frac{a}{(1-az)^2}\right] = \{(n+1)a^{n+1}\}. \qquad (4.195)$$

The relation between the Z-transform and the Laplace transform makes it clear that the Laplace inversion integral can be used for inverting more complicated Z-transforms.

4.10.1 Difference Equations

A sequence $\{u_n\}$ may be characterized by difference equations of the form

$$u_{n+1} = f(u_n), \qquad (4.196)$$

where f is a given function. If f is a linear function, we have a linear difference equation. As an initial condition, u_0 needs to be supplied to begin the evaluation of u_1, u_2, etc. When the equation contains two consequent members, u_{n+1} and u_n, we have a first-order difference equation. In addition to u_{n+1} and u_n, if it contains u_{n+2}, we have a second-order equation. Descretized versions of differential equations (as in numerical analysis) yield difference equations.

4.10.2 First-Order Difference Equation

Consider the equation

$$u_{n+1} = \alpha u_n + \beta. \qquad (4.197)$$

Here α and β are constants. If we denote the Z-transform of $\{u_n\}$ by Z,

$$Z = \mathcal{Z}\{u_n\} = \sum_0^\infty u_n z^n. \qquad (4.198)$$

The Z-transform of the shifted sequence $\{u_{n+1}\}$ can be written as

$$\mathcal{Z}\{u_{n+1}\} = \sum_0^\infty u_{n+1} z^n$$

$$= z^{-1} \sum_0^\infty u_{n+1} z^{n+1}$$

$$= z^{-1} \sum_1^\infty u_n z^n,$$

$$= z^{-1} \left(\sum_0^\infty u_n z^n - u_0 \right)$$

$$= z^{-1}(Z - u_0). \qquad (4.199)$$

With this, we can take the Z-transform of the first-order difference equation, Eq. (4.197). This gives

$$z^{-1}(Z - u_0) = \alpha Z + \frac{\beta}{1-z}, \qquad (4.200)$$

where we have used $\mathcal{Z}\{1\} = 1/(1-z)$. Solving for Z,

$$Z = \frac{u_0}{1-\alpha z} + \frac{z\beta}{(1-z)(1-\alpha z)}. \tag{4.201}$$

Expanding the β-term in partial fractions yields

$$Z = \frac{u_0}{1-\alpha z} + \frac{\beta}{\alpha - 1}\left[\frac{1}{1-\alpha z} - \frac{1}{1-z}\right]. \tag{4.202}$$

Inversion gives

$$u_n = u_0 \alpha^n + \beta \frac{\alpha^n - 1}{\alpha - 1}. \tag{4.203}$$

4.10.3 Second-Order Difference Equation

The second-order homogeneous equation

$$u_{n+2} - (b+c)u_{n+1} + bcu_n = 0, \tag{4.204}$$

with b and c being constants, may be transformed using

$$\mathcal{Z}\{u_n\} = Z, \tag{4.205}$$

$$\mathcal{Z}\{u_{n+1}\} = z^{-1}[Z - u_0], \quad \mathcal{Z}\{u_{n+2}\} = z^{-2}[Z - u_0 - u_1 z], \tag{4.206}$$

to obtain

$$Z - u_0 - u_1 z - (b+c)z[Z - u_0] + bcz^2 Z = 0,$$

$$[1 - (b+c)z + bcz^2]Z = u_0 + u_1 z - (b+c)u_0 z,$$

$$\begin{aligned}
Z &= \frac{u_0 + u_1 z - (b+c)u_0 z}{(1 - bz)(1 - cz)} \\
&= \frac{u_0}{b-c}\left[\frac{b}{1-bz} - \frac{c}{1-cz}\right] + \frac{u_1 - (b+c)u_0}{b-c}\left[\frac{1}{1-bz} - \frac{1}{1-cz}\right] \\
&= \frac{u_1 - cu_0}{b-c}\frac{1}{1-bz} - \frac{u_1 - bu_0}{b-c}\frac{1}{1-cz}.
\end{aligned} \tag{4.207}$$

After inversion, we get

$$u_n = \frac{u_1 - c u_0}{b - c} b^n - \frac{u_1 - b u_0}{b - c} c^n. \tag{4.208}$$

Note that this solution has two unknowns: u_0 and u_1.

Example: Vibration of a String

The motion of a taut string occupying the spatial domain $0 < x < \ell$ is described by

$$T u'' = m \ddot{u}, \tag{4.209}$$

where T is the string tension and m is the mass density per unit length. Assuming harmonic motion in time, we assume

$$u(x,t) = v(x) e^{i\Omega t}, \tag{4.210}$$

to reduce the equation to

$$v'' = -m \Omega^2 v / T. \tag{4.211}$$

Using

$$x/\ell \to x, \quad m\Omega^2 \ell^2 / T = \omega^2, \tag{4.212}$$

we find

$$v'' + \omega^2 v = 0, \quad 0 < x < 1. \tag{4.213}$$

The well-known solution of this equation satisfying the boundary conditions $v(0) = v(1) = 0$ is

$$v(x) = A \sin \omega x, \quad \omega = n\pi. \tag{4.214}$$

Let us approximate the Eq. (4.213) using finite differences by dividing the domain into N equal-length segments. If h is the length of a segment

$$Nh = 1. \tag{4.215}$$

The value of v at an arbitrary point $x_n = nh$ is denoted by v_n and the derivatives are approximated as

$$v' = \frac{v_{n+1} - v_n}{h}, \quad v'' = \frac{v_{n+2} - 2v_{n+1} + v_n}{h^2}. \tag{4.216}$$

The difference form of Eq. (4.213) is

$$v_{n+2} - 2v_{n+1} + (1 + h^2\omega^2)v_n = 0. \tag{4.217}$$

Comparing this with our second-order prototype, Eq. (4.204),

$$b + c = 2, bc = 1 + h^2\omega^2. \tag{4.218}$$

Solving for b and c,

$$b = 1 + ih\omega, \quad c = 1 - ih\omega. \tag{4.219}$$

Substituting in the general solution, Eq. (4.208), with $v_0 = 0$,

$$v_n = v_1 \frac{(1 + ih\omega)^n - (1 - ih\omega)^n}{2ih\omega}. \tag{4.220}$$

To satisfy the condition $v_N = 0$, we require

$$(1 + ih\omega)^N = (1 - ih\omega)^N. \tag{4.221}$$

Taking the Nth root,

$$1 + ih\omega = (1 - ih\omega)e^{2i\pi k/N}, \quad k = 1, 2, \ldots, N - 1. \tag{4.222}$$

Solving for ω,

$$\omega = \frac{1}{ih} \frac{e^{i\pi k/N} - e^{-i\pi k/N}}{e^{i\pi k/N} + e^{-i\pi k/N}}$$

$$= \frac{1}{h} \tan(\pi k/N) = N \tan(\pi k/N). \tag{4.223}$$

For any finite values of nodes N, this equation gives approximate values of the frequency. We may identify k with the mode shape. For large values of N, we may expand this as

$$\omega = N\left[\frac{\pi k}{N} - \frac{\pi^3 k^3}{3N^3} + \cdots\right]$$

$$= \pi k\left[1 - \frac{\pi^2 k^2}{3N^2} + \cdots\right]. \tag{4.224}$$

For the kth mode, the error is given by

$$\frac{\omega - \omega_{exact}}{\omega_{exact}} \approx -\frac{\pi^2 k^2}{3N^2}. \tag{4.225}$$

Of course, as $N \to \infty$, we recover the exact frequency.

4.10.4 Brilluoin Approximation for Crystal Acoustics

To study the propagation of acoustic waves (phonons) in crystals, the physicist, Brilluoin, used an idealized crystal lattice where each atom is considered as a particle of mass m, and the interaction between atoms is approximated using a spring of stiffness k. This is shown in Fig. 4.6.

If the displacement of the nth particle is denoted by $u_n(t)$, its motion is governed by

$$m\ddot{u}_n = k(u_{n+1} + u_{n-1} - 2u_n). \tag{4.226}$$

By changing n to $n+1$, we write this (in our standard form) as

$$u_{n+2} - 2u_{n+1} + u_n = \frac{m}{k}\ddot{u}_{n+1}. \tag{4.227}$$

We assume steady-state motion of the particles in the form

$$u_n(t) = v_n \sin \Omega t, \tag{4.228}$$

to obtain the Helmholtz equation

$$v_{n+2} - (2 - m\Omega^2/k)v_{n+1} + v_n = 0. \tag{4.229}$$

Let

$$\omega = \frac{\Omega}{\sqrt{k/m}}, \tag{4.230}$$

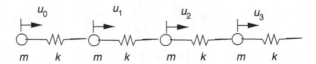

Figure 4.6. A spring-mass representation of a crystal lattice.

where the denominator represents the angular frequency of a simple spring-mass system.

Comparing this with Eq. (4.204),

$$b + c = 2 - \omega^2, \quad bc = 1. \tag{4.231}$$

From this we get

$$b, c = 1 - \frac{\omega^2}{2} \pm \sqrt{\left(1 - \frac{\omega^2}{2}\right)^2 - 1}. \tag{4.232}$$

When $\omega < 2$, we may denote

$$\cos\theta = 1 - \frac{\omega^2}{2}, \tag{4.233}$$

and write

$$b = e^{i\theta}, \quad c = e^{-i\theta}. \tag{4.234}$$

The solution of the Helmholtz equation is

$$\begin{aligned}
v_n &= \frac{v_1 - e^{-i\theta} v_0}{e^{i\theta} - e^{-i\theta}} e^{in\theta} - \frac{v_1 - e^{i\theta} v_0}{e^{i\theta} - e^{-i\theta}} e^{-in\theta} \\
&= v_1 \frac{\sin n\theta}{\sin\theta} - v_0 \frac{\sin(n-1)\theta}{\sin\theta}.
\end{aligned} \tag{4.235}$$

As long as θ is positive real, this solution is bounded for all values of n. When $\omega > 2$, the angle θ is complex, that is the imaginary part of θ is nonzero, and we get unbounded solutions, which cannot be sustained with finite power input. This illustrates the concept of a cut-off frequency,

$$\Omega_{\text{cut-off}} = 2\sqrt{k/m}. \tag{4.236}$$

When Ω is less than the cut-off frequency, we see that our solution is periodic. The significance of a cut-off frequency is that periodic vibration of the system is not possible for frequencies above it.

If n is increased by $2\pi/\theta$, we find

$$v_n = v_{n+N}, \quad N = 2\pi/\theta. \tag{4.237}$$

Here, N is a measure of the wave length, and the relation

$$\theta = 2\pi/N \qquad (4.238)$$

is known as the dispersion relation. From

$$\cos\theta = 1 - \omega^2/2, \qquad (4.239)$$

we get, explicitly,

$$\omega^2 = 2\left[1 - \cos\frac{2\pi}{N}\right]$$
$$\omega = 2\sin\pi/N. \qquad (4.240)$$

Using the periodicity, we may choose $v_0 = 0$ and obtain

$$v_n = v_1 \frac{\sin n\theta}{\sin\theta}. \qquad (4.241)$$

SUGGESTED READING

Andrews, L. C., and Shivamoggi, B. K. (1988). *Integral Transforms for Engineers and Applied Mathematicians*, Macmillan.

Bender, C. M., and Orzag, S. A. (1978). *Advanced Mathematical Methods for Scientists and Engineers*, McGraw-Hill.

Brillouin, L. (1946). *Wave Propagation in Periodic Structures*, McGraw-Hill.

Davies, B. (1985). *Integral Transforms and Their Applications*, 2nd ed., Springer-Verlag.

Ewing, W. M., Jardetzky, W. S., and Press, F. (1957). *Elastic Waves in Layered Media*, McGraw-Hill.

Fung, Y. C. (1965). *Foundations of Solid Mechanics*, Prentice-Hall.

Miles, J. W. (1961). *Integral Transforms in Applied Mathematics*, Cambridge University Press.

Sneddon, I. N. (1972). *The Use of Integral Transforms*, McGraw-Hill.

EXERCISES

4.1 Find the Laplace transforms of

$$f(t) = e^{at}\cos bt, \quad g(t) = e^{at}\sin bt.$$

4.2 Using the relation

$$\int_0^\infty e^{-(a^2x^2+b^2/x^2)}dx = \frac{\sqrt{\pi}}{2a}e^{-2ab}, \quad a,b > 0,$$

obtain the Laplace transforms of

$$f(t) = \frac{1}{\sqrt{t}}e^{-x^2/(4t)}, \quad g(t) = \mathrm{erfc}\left(\frac{x}{2\sqrt{t}}\right).$$

4.3 Find the Laplace transforms of

$$f(t) = a^t, \quad g(t) = \frac{\cos at}{\sqrt{t}},$$

4.4 Invert the transforms

$$\bar{f}(p) = \frac{e^{-2p}}{[(p+1)^2+1]^2}, \quad \bar{g}(p) = \frac{\sqrt{p}}{p-a}.$$

4.5 Show that the Laplace transform of

$$f(t) = \int_1^\infty e^{-t\tau}\frac{d\tau}{\tau}$$

is

$$\bar{f}(p) = \frac{\log(p+1)}{p}.$$

4.6 Invert the transforms

$$\bar{f}(p) = \frac{\cosh ap}{p^3 \cosh p}, \quad \bar{g}(p) = \frac{\sinh ap}{p^2 \sinh p}, \quad 0 < a < 1.$$

4.7 Show that the Laplace transform of the Theta function,

$$\theta(t) = \sum_{n=-\infty}^\infty e^{-n^2\pi^2 t},$$

can be expressed as

$$\bar{\theta}(p) = \frac{\coth\sqrt{p}}{\sqrt{p}}.$$

4.8 Using the expansion

$$\frac{1}{1-e^{-2\sqrt{p}}} = 1 + e^{-2\sqrt{p}} + e^{-4\sqrt{p}} + \cdots ,$$

invert the transform

$$\bar{f}(p) = \frac{\coth\sqrt{p}}{\sqrt{p}}.$$

4.9 Assume $p = a$ is a branch point of the transform $\bar{f}(p)$, and near $p = a$ the transform may be expanded as

$$\bar{f}(p) = c_0(p-a)^{\nu_0} + c_1(p-a)^{\nu_1} + c_2(p-a)^{\nu_2} + \cdots ,$$

where $\nu_0 < \nu_1 < \nu_2 < \cdots$. Show that, as $t \to \infty$, $f(t)$ has the expansion

$$f(t) \sim e^{at}\left[\frac{c_0}{\Gamma(-\nu_0)t^{\nu_0+1}} + \frac{c_1}{\Gamma(-\nu_1)t^{\nu_1+1}} + \frac{c_2}{\Gamma(-\nu_2)t^{\nu_2+1}} + \cdots\right].$$

Use the inversion formula

$$\mathcal{L}^{-1}[p^{\nu}] = \frac{1}{\Gamma(-\nu)t^{\nu+1}}$$

for this purpose.

4.10 Solve

$$\frac{d^2u}{dt^2} + 3\frac{du}{dt} + 2u = e^{2t}\cos t.$$

with $u(0) = du/dt(0) = 0$.

4.11 Solve

$$\frac{d^3u}{dt^3} - 6\frac{d^2u}{dt^2} + 11\frac{du}{dt} - 6u = t,$$

with

$$u(0) = \frac{du}{dt}(0) = \frac{d^2u}{dt^2}(0) = 0.$$

4.12 Solve

$$\frac{d^2u}{dt^2} + 2\frac{du}{dt} + 2u = e^{-t}\sin t, \quad u(0) = \frac{du}{dt}(0) = 0.$$

4.13 Solve the boundary value problem

$$\frac{d^2u}{dx^2} + \frac{du}{dx} - 2u = e^{-x}, \quad u(0) = 1, \quad \frac{du}{dx}(1) = 0.$$

4.14 Solve the system of differential equations

$$\frac{du}{dt} - u + v = 0, \quad \frac{dv}{dt} + v - u = e^t,$$

subject to the initial conditions

$$u(0) = 1, \quad v(0) = 0.$$

4.15 To solve

$$\frac{\partial^2 \psi}{\partial r^2} + \frac{2}{r}\frac{\partial \psi}{\partial r} = \frac{\partial \psi}{\partial t}, \quad \psi(a,t) = T_0, \quad \psi(r,0) = 0,$$

we use a substitution

$$\psi = \phi/r.$$

Obtain the equation for ϕ. Using the Laplace transform solve for ψ.

4.16 Obtain the solution of

$$\frac{\partial^2 u}{\partial x^2} = \frac{\partial^2 u}{\partial t^2}, \quad u(0,t) = U_0 \sin \omega t, \quad u(\ell,t) = 0,$$

$$u(x,0) = \frac{\partial u}{\partial t}(x,0) = 0.$$

4.17 The propagation of elastic stress in a bar: $0 < x < \ell$, is governed by

$$\frac{\partial^2 \sigma}{\partial x^2} = \frac{1}{c^2}\frac{\partial^2 \sigma}{\partial t^2},$$

where c is the wave speed. At time $t = 0$, the end $x = 0$ is subjected to a stress σ_0, while the end $x = \ell$ is subject to $\partial\sigma/\partial x = 0$. The initial conditions are $\sigma(x,0) = 0$ and $\partial\sigma/\partial t(x,0) = 0$. Obtain $\sigma(x,t)$ by expanding $(1 - e^{-2p l/c})^{-1}$ in a binomial series.

4.18 The viscoelastic motion of a semi-infinite bar: $0 < x < \infty$, is governed by

$$c^2\frac{\partial^2 u}{\partial x^2} = \frac{\partial^2 u}{\partial t^2} + \beta\frac{\partial u}{\partial t}.$$

If the boundary and initial conditions are

$$u(0,t) = U_0 \sin \omega t, \quad u(x \to \infty, t) = 0, \quad u(x,0) = \frac{\partial u}{\partial x}(x,0) = 0,$$

show that the solution can be written as

$$u(x,t) = U_0\omega \int_0^t k(x,\tau) \cos \omega(t-\tau)d\tau,$$

where, using a branch cut in the p-plane between $p = 0$ and $p = -\beta$, an expression for $k(x,t)$ can be obtained as

$$k(x,t) = 1 - \frac{1}{\pi} \int_0^\beta \frac{1}{r} e^{-rt} \sin \sqrt{r(\beta-r)}\, dr.$$

4.19 Unsteady heat conduction in a semi-infinite bar is governed by the equation

$$\kappa \frac{\partial^2 u}{\partial x^2} = \frac{\partial u}{\partial t},$$

with the conditions

$$u(x,0) = 0, \quad u(0,t) = U_0 t e^{-at}, \quad u(x \to \infty, t) = 0.$$

Find $u(x,t)$.

4.20 Two semi-infinite bars, A: $-\infty < x < 0$ and B: $0 < x < \infty$ have thermal diffusivities κ_A and κ_B, respectively. Their conductivities are k_A and k_B and they are at uniform temperatures, T_A^0 and T_B^0 when $t < 0$. At time $t = 0$, their ends are made to contact. Obtain the transient temperatures $T_A(x,t)$ and $T_B(x,t)$ for $t > 0$.

4.21 A semi-infinite bar is made of two materials: A and B. Material A occupies $0 < x < 1$ and B occupies $1 < x < \infty$. Heat conduction in the two materials is governed by

$$\kappa_A \frac{\partial^2 T_A}{\partial x^2} = \frac{\partial T_A}{\partial t}, \quad \kappa_B \frac{\partial^2 T_B}{\partial x^2} = \frac{\partial T_B}{\partial t}.$$

The boundary and initial conditions are:

$$T_A(0,t) = T_0 h(t), \quad T_A(x,0) = T_B(x,0) = 0, \quad T_B(x \to \infty) = 0,$$

where $h(t)$ is the Heaviside step function. At the interface $x = 1$,

$$\frac{\partial T_A}{\partial x} = \alpha \frac{\partial T_B}{\partial x}, \quad T_A = T_B,$$

where a is a constant. Obtain the temperatures in the two sections of the bar.

4.22 If u is the solution of

$$\frac{\partial^2 u}{\partial x^2} = \frac{\partial u}{\partial t}, \quad u(0,t) = \frac{\partial u}{\partial x}(0,t) = 0, \quad u(x,0) = f(x),$$

and v is the solution of

$$\frac{\partial^2 v}{\partial x^2} = \frac{\partial^2 v}{\partial t^2}, \quad v(0,t) = \frac{\partial v}{\partial x}(0,t) = 0, \quad v(x,0) = 0, \quad \frac{\partial v}{\partial t}(x,0) = f(x),$$

show that

$$u(x,t) = \frac{1}{\sqrt{4\pi t^3}} \int_0^\infty v(x,\tau) e^{-\tau^2/4t} \tau \, d\tau$$

(from Snedddon, 1972).

4.23 If u is the solution of

$$\frac{\partial^2 u}{\partial x^2} = \frac{\partial u}{\partial t}, \quad u(0,t) = \frac{\partial u}{\partial x}(0,t) = 0, \quad u(x,0) = f(x),$$

and v is the solution of

$$\frac{\partial^2 v}{\partial x^2} = \frac{\partial^2 v}{\partial t^2}, \quad v(0,t) = \frac{\partial v}{\partial x}(0,t) = 0, \quad v(x,0) = f(x), \quad \frac{\partial v}{\partial t}(x,0) = 0,$$

show that

$$u(x,t) = \frac{1}{\sqrt{\pi t}} \int_0^\infty v(x,\tau) e^{-\tau^2/4t} \tau \, d\tau.$$

(from Snedddon, 1972).

4.24 Solve the integral equation

$$u(t) - a \int_0^t e^{-a\tau} u(t-\tau) d\tau = e^{-bt},$$

where a and b are constants.

4.25 Solve the integral equation

$$\int_0^t \frac{\cos[k\sqrt{t^2 - \tau^2}]}{\sqrt{t^2 - \tau^2}} f(\tau) d\tau = g(t).$$

4.26 Solve the integral equation

$$\int_0^x \frac{u(\xi)d\xi}{(x - \xi)^{2/3}} = \int_0^x \frac{f(\xi)d\xi}{(x - \xi)^{1/3}}.$$

4.27 Solve the integral equation

$$\int_0^x \frac{u(\xi)d\xi}{\sqrt{x^2 - \xi^2}} = \frac{1}{\sqrt{x}}.$$

4.28 Find the solution of

$$u_{n+1} - au_n = nb^n, \quad u_0 = 1.$$

4.29 Find the solution of the difference equation

$$u_{n+2} - 2bu_{n+1} + b^2 u_n = b^n, \quad u_0 = u_1 = 0.$$

4.30 Obtain the frequencies ω corresponding to the periodic solutions of

$$u_{n+2} - (2 - \omega^2)u_{n+1} + u_n = 0,$$

subject to the conditions

$$u_0 = u_4 = 1, \quad u_2 = u_5.$$

Author Index

Subject Index

Printed in the United States
By Bookmasters